introductory statistics
for the behavioral sciences

introductory statistics for the behavioral sciences

JOAN WELKOWITZ

ROBERT B. EWEN

JACOB COHEN

New York University

Academic Press New York and London

ACADEMIC PRESS, INC.
111 Fifth Avenue, New York, New York
10003

United Kingdom Edition published by
ACADEMIC PRESS, INC. (LONDON) LTD.
Berkeley Square House, London W1X 6BA

Library of Congress Catalog Card Number: 76-152747

Printed in the United States of America

Dedicated to
  our students—past, present, and future—
  to Julie, Larry, and David
  and to Pat and Erika

contents

Preface

This book represents the efforts of three authors who have jointly accumulated many years of experience in statistical procedures through teaching and research efforts. Our purpose has been to introduce and explain statistical concepts and principles clearly and in a highly readable fashion, assuming minimal mathematical sophistication but at the same time avoiding a "cookbook" approach to methodology.

We have attempted to present a broader outlook on hypothesis testing than is customary by devoting an entire chapter to the much neglected concepts of statistical power and the probability of a Type II error. To our knowledge, this is the first time that power tables that can easily be used by beginning students of statistics have been included in an introductory statistics textbook. As another important extension of conventional tests of significance, the conversion of t values and other such results of significance tests to correlational-type indices which express *strength* of relationship have also been included. Also, special time-saving procedures for hand calculation which have become outmoded by the ready availability of electronic calculators and computers, such as the computation of means and standard deviations from grouped frequency distributions, have been omitted.

Throughout the text, the robustness of parametric procedures has been emphasized. That is, such significance tests as t and F yield accurate results even if such assumptions as equal population variances and normal population distributions are not well met. Therefore, nonparametric procedures (with the exception of chi square) have not in general been included.

The text is accompanied by a workbook of problems which summarizes important procedures at the beginning of each chapter, and affords the opportunity for practice in the computational procedures and practice in the *choice* of the correct procedure and understanding of the underlying rationale.

Acknowledgments

Thanks are due to the many encouraging friends and relatives; to the infinitely patient staff of Academic Press; and most of all to our students who have provided invaluable feedback to both the textbook chapters and the workbook.

We also wish to thank Dr. Joseph L. Fleiss of Columbia University and Dr. Harold Fletcher of Florida State University for their most helpful comments and suggestions. We are also indebted to the Literary Executor of the late Sir Ronald A. Fisher, F.R.S., to Dr. Frank Yates, F.R.S., and to Oliver & Boyd, Edinburgh, for permission to reprint Tables C, D, and G from their book *Statistical Tables for Biological, Agricultural, and Medical Research*, and to the Iowa State University Press, Drs. J. W. Dunlap and A. K. Kurtz, the Houghton Mifflin Company and *Psychometrika*, Dr. A. L. Edwards, the Holt, Rinehart, and Winston Company, and the *Annals of Mathematical Statistics* for permission to reproduce the other statistical tables in the Appendix.

glossary of symbols

Numbers in parentheses indicate the chapter in which the symbol first appears.

a	Y intercept of line in linear regression (11)
α	criterion (or level) of significance; probability of Type I error (9)
b_{YX}	slope of linear regression line for predicting Y from X (11)
b_{XY}	slope of linear regression line for predicting X from Y (11)
β	probability of Type II error (9)
$1 - \beta$	power (13)
C	contingency coefficient (16)
cf	cumulative frequency (2)
χ^2	chi square (16)
D	difference between two scores or ranks (10)
$\bar{D}$	mean of the Ds (10)
df	degrees of freedom (9)
df_B	degrees of freedom between groups (14)
df_W	degrees of freedom within groups (14)
df_1	degrees of freedom for factor 1 (15)
df_2	degrees of freedom for factor 2 (15)
$df_{1 \times 2}$	degrees of freedom for interaction (15)
δ	delta (13)
ε	epsilon (14)
f	frequency (2)
f_e	expected frequency (16)
f_o	observed frequency (16)
F	statistic following the F distribution (14)
γ	effect size, gamma (13)
h	interval size (3)
$H\%$	percent of subjects in all intervals higher than the critical one (3)
H_0	null hypothesis (9)
H_1	alternative hypothesis (9)
i	case number (1)
$I\%$	percent of subjects in the critical interval (3)
k	a constant (1)
k	number of groups (or the last group) (14)
$L\%$	percent of subjects in all intervals below the critical one (3)
LRL	lower real limit (3)
Mdn	median (4)
MS	mean square (14)
MS_B	mean square between groups (14)
MS_W	mean square within groups (14)
MS_1	mean square for factor 1 (15)

MS_2	mean square for factor 2 (15)
$MS_{1 \times 2}$	mean square for interaction (15)
μ	population mean (4)
N	number of subjects or observations (1)
N_G	number of observations or subjects in group G (14)
π	hypothetical population proportion (9)
p	observed sample proportion (9)
$P(A)$	probability of event A (7)
PR	percentile rank (3)
ϕ	phi coefficient (16)
r_{XY}	sample Pearson correlation coefficient between X and Y (11)
r_s	Spearman rank-order correlation coefficient (12)
r_{pb}	point-biserial correlation coefficient (12)
ρ_{XY}	population correlation coefficient between X and Y (11)
s	sample standard deviation (5)
s^2	population variance estimate (5)
$s_D{}^2$	variance of the Ds (10)
s^2_{pooled}	pooled variance (10)
$s_{\bar{X}}$	standard error of the mean (9)
$s_{\bar{X}_1 - \bar{X}_2}$	standard error of the difference (10)
$Score_p$	score corresponding to the pth percentile (3)
SFB	sum of frequencies below the critical interval (3)
SS	sum of squares (14)
SS_T	total sum of squares (14)
SS_B	sum of squares between groups (14)
SS_W	sum of squares within groups (14)
SS_1	sum of squares for factor 1 (15)
SS_2	sum of squares for factor 2 (15)
$SS_{1 \times 2}$	sum of squares for interaction (15)
$\sum$	sum or add up (1)
σ	standard deviation (5)
σ^2	variance (5)
$\sigma_{\bar{X}}$	standard error of the mean when σ is known (9)
σ_p	standard error of a sample proportion (9)
$\sigma_{Y'}$	standard error of estimate for predicting Y (11)
$\sigma_{X'}$	standard error of estimate for predicting X (11)
t	statistic following the t distribution (9)
T	T score (6)
θ	theta (13)
x	deviation score (4)
X'	predicted X score (11)
$\bar{X}$	sample mean (4)
$\bar{X}_G$	mean of group G (14)
Y'	predicted Y score (11)
Z	standard score (6)
z	standard score based on a normal distribution (8)

part **I** introduction

1 introduction

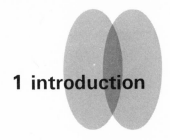

This book is written primarily for undergraduates majoring in psychology, sociology, and education. There are three major reasons why a knowledge of statistics is essential for those who wish to pursue the study of these behavioral sciences:

To understand the professional literature. The vast majority of professional literature in the behavioral sciences includes results that are based on statistical analyses. Regardless of what areas are of particular interest to you, you will be unable to understand many relevant articles in scientific journals and books unless you understand statistics. It is possible to seek out second-hand reports that have been carefully edited in order to be comprehensible to the statistically ignorant, but those who prefer this alternative to obtaining first-hand information should not be majoring in the fields of behavioral science.

To understand the rationale underlying research in the behavioral sciences. Statistics is not just a catalog of procedures and formulas; it offers the rationale upon which much of behavioral science research is based—namely, drawing inferences about a population based on data obtained from a sample. Those familiar with statistics readily understand this concept and recognize that research consists of a series of educated guesses and fallible decisions, and not right or wrong answers. Those without a knowledge of statistics, on the other hand, invariably fail to comprehend the essential nature of the strengths and weaknesses of the techniques by which behavioral scientists collect information and draw conclusions.

To carry out behavioral science research. In order to do competent research in the behavioral sciences, it is necessary to design the statistical analysis *before* the data are collected. Otherwise, the research procedures may be so poorly planned that not even an expert statistician can make any sense out of the results. To be sure, it is possible (and often advisable) to consult someone more experienced in statistics for assistance. Without some statistical knowledge of your own, however, you will find it difficult or impossible to convey your needs to someone else and to understand his replies.

Save for these introductory remarks, we do not regard it as our task to persuade you that statistics is important in psychology, sociology, and education. If you are seriously interested in the study of any of these fields, you will find this out for yourself (if you have not already done so). Accordingly, this book contains no documented examples selected from the professional literature to prove to you that statistics really is used in these fields. Instead, we have used realistic examples with numerical values chosen so as to reveal the issues involved as clearly as possible.

In addition, we have tried to avoid a "cookbook" approach which places excessive emphasis on computational recipes. Instead, the reason for the various statistical procedures and the essential underlying concepts have been explained at length, and insofar as possible in standard English, so that you will know not only what to do but why you are doing it. Do not, however, expect to learn the material in this book from a single reading. The concepts involved in statistics, especially inferential statistics, are sufficiently challenging so that it is often said that the only way to completely understand statistics is to teach it (or write a book about it). On the other hand, there is no reason to approach statistics with fear and trembling. You do not have to be a mathematical expert to obtain a good working knowledge of statistics; what *is* needed is mathematical comprehension sufficient to cope with high-school algebra and a willingness to work at new concepts until they are understood.

DESCRIPTIVE AND INFERENTIAL STATISTICS

One purpose of statistics is to summarize or describe the characteristics of a set of data in a clear and convenient fashion; this is accomplished by what are called *descriptive statistics*. You are already familiar with some descriptive statistics; for example, your grade-point average serves as a convenient summary of all of the grades that you have received in college. Part 2 of this book is devoted to descriptive statistics.

A second function of statistics is to make possible the solution of an extremely important problem. Behavioral scientists can never measure *all* of the cases in which they are interested. For example, a clinical psychologist studying the effects of various kinds of therapies cannot obtain data on all of the mental patients in the world; a social psychologist studying ethnic differences in attitudes cannot measure all of the millions of blacks and whites in the United States; an experimental psychologist cannot observe the maze behavior of all rats. The behavioral scientist wants to know what is happening in a given *population*—a large group (theoretically an infinitely large group) of people, animals, objects, or responses that are alike in at least one respect (for example, all blacks in the United States). He cannot measure the entire population, however, because it is so large that it would be much too time-consuming and expensive to do so. What is he to do?

One reasonable procedure is to measure a relatively small number of cases drawn from the population (that is, a *sample*). Thus, a sample of 100 people could readily be interviewed or given a written questionnaire. However, conclusions that apply only to the 100 people who happened to be included in the sample are unlikely to be of much interest. The behavioral scientist hopes to advance scientific knowledge by drawing general conclusions— for example, about all blacks and all whites in the populations from which the samples come and not just the 50 blacks and 50 whites in the samples. *Inferential statistics* makes it possible to draw inferences about what is happening in the population based on what is observed in a sample from that population. (This point will be discussed at greater length in Chapter 7.) Part 3 of this book is devoted to inferential statistics (which make frequent use of some of the descriptive statistics discussed in Part 2).

POPULATIONS, SAMPLES, PARAMETERS, AND STATISTICS

As the above discussion indicates, the term *population* as used in statistics does not necessarily refer to people. For example, the population of interest might be that of all white rats of a given genetic strain, or all responses of a single subject's eyelid in a conditioning experiment. Whereas the population consists of all of the cases of interest, a *sample* consists of any subgroup drawn from the specified population. It is important that the population be clearly specified; for example, a group of 100 New York University freshmen might be a well-drawn sample from the population of all

NYU freshmen or a poorly drawn sample from the population of all undergraduates in the United States. It is strictly proper to apply (that is, *generalize*) the research results only to the specified population. (A researcher *may* justifiably argue that his results are more widely generalizable, but he is on his own if he does so because the rules of statistical inference as such do not justify this.)

A *statistic* is a numerical quantity (such as an average) which summarizes some characteristic of a sample. A *parameter* is the corresponding value of that characteristic in the population. For example, if the average studying time of a sample of 100 NYU freshmen is 7.4 hours per week, then 7.4 is a statistic. If the average studying time of the population of all NYU freshmen is 9.6 hours per week, then 9.6 is the corresponding population parameter. Usually the values of population parameters are unknown because the population is too large to measure in its entirety, and appropriate techniques of inferential statistics are used to estimate the values of population parameters from sample statistics. If the sample is properly selected, the sample statistics will often give good estimates of the parameters of the population from which the sample was drawn; if the sample is poorly chosen, erroneous conclusions are likely to occur. Thus, whether you are doing your own research or reading about that of someone else, you should always check to be sure that the population to which the results are generalized is proper in light of the sample from which the results were obtained.

SUMMATION NOTATION

Mathematical formulas and symbols often appear forbidding. In fact, however, they save time and trouble by clearly and concisely conveying information that would be very awkward to express in words. In statistics, a particularly important symbol is the one used to represent the *sum* of a set of numbers (the value obtained by adding up the numbers).

To illustrate the use of summation notation, let us suppose that eight students take a ten-point quiz. Letting X stand for the variable in question (quiz scores), let us further suppose that the results are as follows:

$$X_1 = 7 \quad X_2 = 9 \quad X_3 = 6 \quad X_4 = 10$$
$$X_5 = 6 \quad X_6 = 5 \quad X_7 = 3 \quad X_8 = X_N = 4$$

Notice that X_1 represents the first score on X; X_2 stands for the second score on X; and so on. Also, the *number of scores* is denoted by N; in this example, $N = 8$. The last score may be represented either by X_8 or X_N. The *sum of all the X scores* is represented by

$$\sum_{i=1}^{N} X_i$$

where $\sum$, the Greek capital letter sigma, stands for "the sum of" and is called the *summation sign*. The subscript below the summation sign indicates that the sum begins with the first score (X_i where $i = 1$), and the superscript above the summation sign indicates that the sum continues up to and including the last score (X_i where $i = N$ or 8). Thus,

$$\sum_{i=1}^{N} X_i = X_1 + X_2 + X_3 + X_4 + X_5 + X_6 + X_7 + X_8$$
$$= 7 + 9 + 6 + 10 + 6 + 5 + 3 + 4$$
$$= 50$$

In some instances, the sum of only a subgroup of the numbers may be needed. For example, the symbol

$$\sum_{i=3}^{6} X_i$$

represents the sum beginning with the third score (X_i where $i = 3$) and ending with the sixth score (X_i where $i = 6$). Thus,

$$\sum_{i=3}^{6} X_i = X_3 + X_4 + X_5 + X_6$$
$$= 6 + 10 + 6 + 5$$
$$= 27$$

Most of the time, however, the sum of *all* the scores is needed in the statistical analysis, and in such situations it is customary to omit the indices i and N from the notation, as follows:

$$\sum X = \text{sum of all the } X \text{ scores}$$

The fact that there is no written indication as to where to begin and end the summation is taken to mean that all the X scores are to be summed.

SUMMATION RULES

Certain rules involving summation notation will prove useful in subsequent chapters. Let us suppose that the eight students mentioned previously take a second quiz, denoted by Y. The results of both quizzes can be summarized conveniently as follows:

subject (S)	quiz 1 (X)	quiz 2 (Y)
1	7	8
2	9	6
3	6	4
4	10	10
5	6	5
6	5	10
7	3	9
8	4	8

We have already seen that $\sum X = 50$. The sum of the scores on the second quiz is equal to:

$$\sum Y = Y_1 + Y_2 + Y_3 + Y_4 + Y_5 + Y_6 + Y_7 + Y_8$$
$$= 8 + 6 + 4 + 10 + 5 + 10 + 9 + 8$$
$$= 60$$

The following rules are illustrated using the small set of data shown above, and you should verify each one carefully.

Rule 1. $\sum (X + Y) = \sum X + \sum Y$.

ILLUSTRATION S	X	Y	X + Y
1	7	8	15
2	9	6	15
3	6	4	10
4	10	10	20
5	6	5	11
6	5	10	15
7	3	9	12
8	4	8	12

$$\sum X = 50 \quad \sum Y = 60 \quad \sum (X + Y) = 110$$
$$\sum X + \sum Y = 110$$

This rule should be intuitively obvious; the same total should be reached regardless of the order in which the scores are added.

Rule 2. $\sum (X - Y) = \sum X - \sum Y$.

ILLUSTRATION S	X	Y	X - Y
1	7	8	-1
2	9	6	3
3	6	4	2
4	10	10	0
5	6	5	1
6	5	10	-5
7	3	9	-6
8	4	8	-4

$$\sum X = 50 \quad \sum Y = 60 \quad \sum (X - Y) = -10$$
$$\sum X - \sum Y = -10$$

Similar to the first rule, it makes no difference whether you subtract first and then sum $[\sum (X - Y)]$ or obtain the sums of X and Y first and then subtract $(\sum X - \sum Y)$.

Unfortunately, matters are not so simple when multiplication and squaring are involved:

Rule 3. $\sum XY \neq \sum X \sum Y$. That is, first multiplying each X by the corresponding Y (XY) and then summing $(\sum XY)$ is *not* equal to summing X ($\sum X$) and summing Y ($\sum Y$) first and then multiplying once $(\sum X \sum Y)$.

ILLUSTRATION S	X	Y	XY
1	7	8	56
2	9	6	54
3	6	4	24
4	10	10	100
5	6	5	30
6	5	10	50
7	3	9	27
8	4	8	32

$$\sum X = 50 \quad \sum Y = 60 \quad \sum XY = 373$$
$$\sum X \sum Y = (50)(60) = 3000$$

Observe that $\sum XY = 373$, while $\sum X \sum Y = 3000$.

Rule 4. $\sum X^2 \neq (\sum X)^2$. That is, first squaring all X values and then summing $(\sum X^2)$ is *not* equal to summing first and then squaring a single quantity $[(\sum X)^2]$.

ILLUSTRATION S	X	X²
1	7	49
2	9	81
3	6	36
4	10	100
5	6	36
6	5	25
7	3	9
8	4	16

$$\sum X = 50 \qquad \sum X^2 = 352$$
$$(\sum X)^2 = (50)^2 = 2500$$

Here, $\sum X^2 = 352$, while $(\sum X)^2 = 2500$.

Rule 5. If k is a *constant* (a fixed numerical value), then $\sum k = Nk$.

ILLUSTRATION   Suppose that $k = 3$. Then,

S	k
1	3
2	3
3	3
4	3
5	3
6	3
7	3
8	3

$$\sum k = 24$$
$$Nk = (8)(3) = 24$$

Rule 6. If k is a constant,

$$\sum (X + k) = \sum X + \sum k = \sum X + Nk.$$

ILLUSTRATION   Suppose that $k = 5$. Then,

S	X	k	X + k
1	7	5	12
2	9	5	14
3	6	5	11
4	10	5	15
5	6	5	11
6	5	5	10
7	3	5	8
8	4	5	9

$$\sum X = 50 \quad \sum k = Nk = 40 \quad \sum (X + k) = 90$$
$$\sum X + Nk = 50 + (8)(5) = 90$$

This rule follows directly from Rules 1 and 5.

Rule 7. If k is a constant,

$$\sum (X - k) = \sum X - Nk.$$

The illustration of this rule is similar to that of Rule 6 and is left to the reader as an exercise.

Rule 8. If k is a constant,

$$\sum kX = k \sum X.$$

ILLUSTRATION   Suppose that $k = 2$. Then,

S	X	k	kX
1	7	2	14
2	9	2	18
3	6	2	12
4	10	2	20
5	6	2	12
6	5	2	10
7	3	2	6
8	4	2	8
	$\sum X = 50$		$\sum kX = 100$

$$k \sum X = (2)(50) = 100$$

SUMMARY

Descriptive statistics are used to summarize and make understandable large quantities of data. *Inferential statistics* are used to draw inferences about numerical quantities (called *parameters*) concerning *populations* based on numerical quantities (called *statistics*) obtained from *samples*. The *summation sign*, $\sum$, is used to indicate "the sum of" and occurs frequently in statistical work.

part **II** descriptive statistics

2 frequency distributions and graphs

THE PURPOSE OF DESCRIPTIVE STATISTICS

The primary goal of descriptive statistics is to bring order out of chaos. For example, consider the plight of a professor who has given an examination to a large class of 85 students and has computed the total score on each student's exam. In order to decide what represents good and bad performance on the examination, the professor must find a way to comprehend and interpret 85 numbers (test scores), which can be a very confusing task. Similarly, a researcher who runs 60 rats through a maze and records the time taken to run the maze on each trial is faced with the problem of interpreting 60 separate numbers.

In addition to causing problems for the professor or researcher, the large quantity of numbers also creates difficulties for the audience in question. The students in the first example are likely to request the distribution of test scores so as to be able to interpret their own performance, and they also will have trouble trying to interpret 85 unorganized numbers. Likewise, the people who read the scientific paper ultimately published by the researcher interested in rats and mazes will have a highly traumatic time trying to interpret a table with 60 numbers in it.

Descriptive statistics help resolve problems such as these by making it possible to *summarize and describe large quantities of data.* Among the various techniques that you will find particularly useful are the following:

Frequency distributions and graphs—procedures for describing all (or nearly all) of the data in a convenient way.

Measures of "central tendency"—single numbers that describe the location of a distribution of scores, that is, where it generally falls within the infinite range of possible values.

Measures of variability—single numbers that describe how "spread out" a set of scores is (whether the numbers are similar to each other and hence vary very little, as opposed to whether they tend to be very different from one another and hence vary a great deal).

Transformed scores—new scores which replace each original number and show at a glance how good or bad any score is in comparison to the other scores in the group.

Each of these procedures serves a different (and important) function. Our discussion of descriptive statistics will begin with frequency distributions and graphs; the other topics will be treated in subsequent chapters.

REGULAR FREQUENCY DISTRIBUTIONS

One way of making a set of data more comprehensible is to write down every possible score value in order, and next to each score value record the number of times that the score occurs. For example, suppose that an experimenter interested in human ability obtains a group of 100 college undergraduates and gives each one 10 problems to solve; a subject's score may therefore fall between zero (none correct) and 10 (all correct), inclusive. The scores of the 100 subjects are shown in Table 2.1.

As you can see, the table of 100 numbers is rather confusing and difficult to comprehend. A *regular frequency distribution*, on the other hand, will present a clearer picture. The first step in constructing such a distribution is to list every *score value* in the first column

Table 2.1 *Number of problems solved correctly by 100 college undergraduates* (*hypothetical data*)

5	8	3	6	5	8	3	7	4	7
6	8	7	4	6	5	8	7	10	6
5	3	6	1	8	6	8	5	7	10
8	9	6	9	6	5	9	9	6	3
8	5	8	4	8	6	7	4	10	5
8	7	6	8	5	7	6	9	6	10
4	8	6	8	6	5	4	7	9	6
7	4	5	5	9	6	6	7	4	5
10	3	5	7	9	10	6	7	6	6
6	9	8	7	8	5	7	4	6	8

of a table (frequently denoted by the symbol X), with the highest score at the top. Second, the *frequency* (denoted by the symbol f) of each score, or the number of times a given score was obtained, is listed to the right of the score in the second column of the table. To arrive at the figures in the "frequency" column, you could go through the data and count all the tens, go through the data again and count all the nines, and so forth, until all frequencies were tabulated. A more efficient plan, however, is to go through the data just once and make a tally mark next to the appropriate score in the score column for each score, and add up the tally marks at the end. The completed regular frequency distribution is shown in Table 2.2

Table 2.2 *Regular and cumulative frequency distributions for data in Table 2.1*

score (X)	frequency (f)	cumulative frequency (cf)
10	6	100
9	9	94
8	17	85
7	15	68
6	23	53
5	15	30
4	9	15
3	5	6
2	0	1
1	1	1
0	0	0

(ignore the "cumulative frequency" column for the moment; it will be discussed in the next section). The table reveals at a glance how often each score was obtained; for example, nine people received a score of 9 and five people received a score of 3. This makes it easier to form impressions as to the performance of subjects in the experiment, since you can conveniently ascertain (among other things) that 6 was the most frequently obtained score, scores distant from 6 tended to occur less frequently than scores close to 6, and the majority of people got more than half the problems correct.

CUMULATIVE FREQUENCY DISTRIBUTIONS

The primary value of *cumulative frequency distributions* will not become apparent until subsequent chapters, when they will prove to be of assistance in the computation of certain statistics (such as the median and percentiles). To construct a cumulative frequency distribution, first form a regular frequency distribution. Then, start with the *lowest* score in the distribution and form a new column of *cumulative frequencies* by adding up the frequencies as you go along. For example, the following diagram shows how the cumulative frequencies (denoted by the symbol *cf*) in the lower portion of Table 2.2 were obtained:

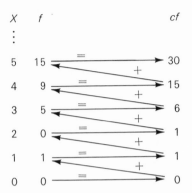

That is, the cumulative frequency for the score of 4 is equal to 15. This value was obtained by adding the frequency (*f*) of that score, namely 9, to the cumulative frequency (*cf*) of the next lower score, which is 6. The other cumulative frequencies were obtained in a similar fashion.

The cumulative frequency distribution is interpreted as follows: the cumulative frequency of 15 for the score of 4 means that 15 people obtained a score of *4 or less*. This can be readily verified by looking at the frequencies (*f*): nine people scored 4; five people scored 3, and one person scored 1, for a total of 15 people with scores at or below 4:

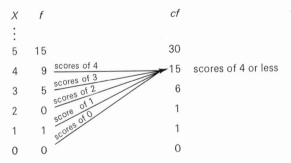

The *cf* for the score of 7 in Table 2.2 is 68, which means that 68 people obtained a score of *7 or less*. Since everyone obtained a score of *10 or less*, the *cf* for a score of 10 should equal the total number of people (100) if the cumulative frequencies are correct—

which it does. (The *cf* for the *highest* score should always equal *N*, the number of people.) With this experience under your belt, can you tell how many people obtained a score of *6 or more?* The cumulative frequency distribution reveals that 30 people obtained scores of 5 or less, so there must be *70* people who scored 6 or more. This can be checked by using the frequencies; there were *23* scores of 6 + *15* scores of 7 + *17* scores of 8 + *9* scores of 9 + *6* scores of 10. Although no unique information is presented by the cumulative frequency distribution, it allows you to arrive at certain needed information more quickly and conveniently than is possible using the regular frequency distribution.

GROUPED FREQUENCY DISTRIBUTIONS

When the number of *different scores* to be listed in the score (*X*) column is not too large, regular frequency distributions are an excellent way of conveniently summarizing a set of data. If, however, there are more than 15 or 20 values of *X* to be written, constructing a regular frequency distribution is likely to prove very tedious. One way of avoiding an excessive case of writer's cramp and a summary that does not summarize enough is to use a *grouped frequency distribution*. Instead of listing single scores in the score column (for example, 0, 1, 2, 3, 4, 5, 6, 7,...), several score values are grouped together into a *class interval* (for example, 0–4, 5–9, 10–14, 15–19,...), and frequencies are tallied for each interval.

For example, the data in Table 2.3 represent hypothetical scores of 85 students on a 50-point midterm examination. Inspection of the data shows that the scores range from a low of 9 to a high of 49. Therefore, if a regular frequency distribution were to be used, some

Table 2.3 *Scores of 85 students on a 50-point midterm examination (hypothetical data)*

39	42	30	11	35	25	18	26	37	· 15
29	22	33	32	21	43	11	11	32	29
44	26	30	49	13	38	26	30	45	21
31	28	14	35	10	41	15	39	33	34
46	21	38	26	26	37	37	14	26	24
32	15	22	28	33	47	9	22	31	20
37	40	20	39	30	18	29	35	41	21
26	25	29	33	23	30	43	28	32	32
34	28	38	32	31					

41 separate scores and corresponding frequencies would have to be listed.* To avoid such a tiresome task, a grouped frequency distribution has been formed in Table 2.4. An *interval size* of 3 has

Table 2.4 *Grouped and cumulative frequency distributions for data in Table 2.3*

class interval	frequency (f)	cumulative frequency (cf)
48–50	1	85
45–47	3	84
42–44	4	81
39–41	6	77
36–38	7	71
33–35	9	64
30–32	14	55
27–29	8	41
24–26	10	33
21–23	8	23
18–20	4	15
15–17	3	11
12–14	3	8
9–11	5	5

been chosen, meaning that there are three score values in each class interval. (The symbol h will be used to denote interval size.) Then, successive intervals of size 3 are formed until the entire range of scores has been covered. Next, frequencies are tabulated, with all scores falling in the same interval being treated equally. For example, a score of 39, 40, or 41 would be entered by registering a tally mark next to the class interval 39–41. When the tabulation is completed, the frequency opposite a given class interval indicates the number of cases with scores in that interval. (Thus, grouped frequency distributions lose information, since they do not provide the exact value of each score. They are very convenient for purposes of summarizing a set of data, but should not generally be used when computing means and other statistics.)

In the illustrative problem, the interval size of 3 was specified. In your own work, there will be no such instructions, and it will be up to you to construct the proper intervals. The conventional procedure is to select the intervals in such a way as to satisfy the following guidelines:

* There are 49 − 9 + 1 or 41 numbers between 9 and 49, inclusive.

1. Have a total of from 8 to 15 class intervals.
2. Use an interval size of 2, 3, 5, or a multiple of 5, selecting the smallest size that will satisfy the first rule. (All intervals should be the same size.)
3. Make the lowest score in each interval a multiple of the interval size.

For example, suppose that scores range from a low of 49 to a high of 68. There are a total of 20 score values ($68 - 49 + 1$), and an interval size of 2 will yield 20/2 or 10 class intervals. This falls within the recommended limits of 8 and 15, and size 2 should therefore be selected. It would waste far too much time and effort to list all the nonoccurring scores between zero and 48, and the first interval should be 48–49, the next interval 50–51, and so on, so that the first score in each interval will be evenly divisible by the interval size, 2. If instead scores range from a low of zero to a high of 52, there are 53 score values in all. An interval size of 2 or 3 will yield too many intervals ($53/2 = 26+$; $53/3 = 17+$), and size 4 is customarily avoided (being so close to 5, which produces intervals more like the familiar decimal system). Therefore, interval size 5, which will produce 11 intervals, should be chosen; begin with the class interval 0–4 and continue with 5–9, 10–14, 15–19, and so on. What if scores range from 14 to 25? In this case, there are only 12 possible score values and you *should not group* the data, because not enough time and effort will be saved to warrant the loss of information. Use a regular frequency distribution.

Just as was the case with regular frequency distributions, a cumulative frequency distribution can be formed from grouped data, and you will find one in Table 2.4. The cumulative frequencies are formed by starting with the lowest interval and adding up the frequencies as you go along, and are interpreted in the usual way; for example, the value of 64 corresponding to the class interval 33–35 means that 64 people obtained scores at or below this interval—or, in other words, scores of 35 or less.

GRAPHIC REPRESENTATIONS

It is often effective to express frequency distributions pictorially as well as in tables. Two procedures for accomplishing this are discussed below.

HISTOGRAMS

Suppose that a sample of 25 families is obtained and the number of children in each family is recorded, and the results are as follows:

number of children (X)	f
7 or more	0
6	1
5	0
4	3
3	4
2	8
1	5
0	4

A *histogram* (or "bar graph") of these data is shown in Figure 2.1. To construct the histogram, the Y axis (vertical axis) is marked off in terms of *frequencies*, and the X axis (horizontal axis) is marked off in terms of *score values*. In each case, a little forethought (and inspection of the total number of score values and largest single frequency in the frequency distribution) will prevent such disasters as having to enter several large frequencies or scores on the desk instead of on the histogram, or winding up with a histogram that can be seen only through a microscope. The frequency of any score is expressed by the height of the bar above that score; thus, to show that a score of 3 (children) occurred 4 times, a bar 4 units in

Figure 2.1 *Histogram expressing number of children per family in a sample of 25 families (hypothetical data).*

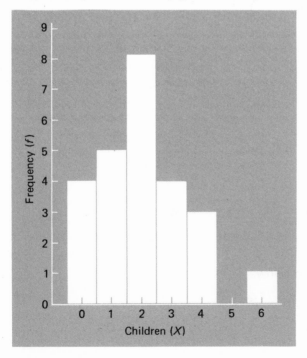

height is drawn above this score. To construct a histogram from a grouped frequency distribution, simply use the midpoint of each class interval (the average of the highest and lowest scores in the interval) * as the score value for that interval.

Histograms can be used for any kind of data but are particularly appropriate for *discrete* data, where results between the score values shown *cannot* occur. In the present example, it is impossible for a family to have 2.4 children, and this fact is well expressed in the histogram by the separate and distinct bars above each score value.

FREQUENCY POLYGONS

Regular frequency polygons. The frequency data in Table 2.4 are presented as a *regular frequency polygon* in Figure 2.2. The strategy is the same as in the case of the histogram, in that the frequency of a score is expressed by the height of the graph above the X axis, and frequencies are entered on the Y axis and scores on the X axis. The difference is that points, rather than bars, are used for each entry. Thus, the frequency of 5 for the class interval of 9–11 is shown by a dot 5 units up on the Y axis above the score of 10 (the midpoint of the 9–11 interval; with a regular frequency

Figure 2.2   *Regular frequency polygon for data in Table 2.4.*

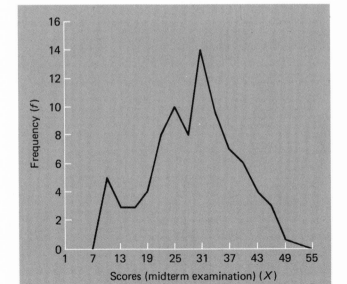

* Note that age, when taken as of one's last birthday, would be an exception to this rule. For example, the midpoint for 10-year-olds would be $10\frac{1}{2}$ years.

distribution, each score would be entered on the X axis). All dots are connected with straight lines, and the resulting frequency polygon provides a pictorial illustration of the frequency distribution.

Regular frequency polygons are particularly appropriate for *continuous* data, where results between the score values shown can occur, or could if it were possible to measure with sufficient refinement. Scores, whether or not fractions can occur, are always treated as continuous.

Cumulative frequency polygons. Cumulative frequency distributions are also commonly graphed in the form of frequency polygons, and the resulting figure is called (not very surprisingly) a *cumulative frequency polygon.* An example, based on the data in Table 2.2 is provided in Figure 2.3. Note that, reading from left to right, the cumulative frequency distribution always remains level or increases and can never drop down toward the X axis; this is because the cumulative frequencies are formed by successive additions and the *cf* for an interval can be at most equal to, but never less than, the *cf* for the preceding interval.

Figure 2.3 *Cumulative frequency polygon for data in Table 2.2.*

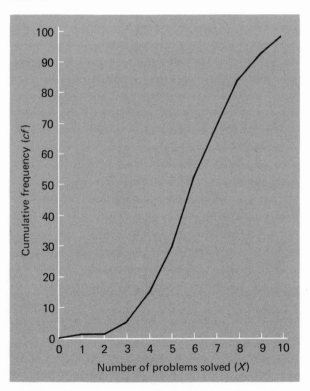

SHAPES OF FREQUENCY DISTRIBUTIONS

It is often useful to talk about the general shape of a frequency distribution. Some important definitions commonly used in this regard are as follows:

Symmetry versus skewness. A distribution is *symmetric* if and only if it can be divided into two halves, each the "mirror image" of the other. For example, distributions A, B, C, and G in Figure 2.4 are symmetric. Distributions D, E, F, and H are *not* symmetric, however, since they cannot be divided into two similar parts.

A markedly asymmetric distribution with a pronounced "tail" is described as *skewed* in the direction of the tail. Thus, Distribution E is *skewed to the left* (or negatively skewed), since the long tail

Figure 2.4   *Shapes of frequency distributions. (A) Normal curve (symmetric, unimodal). (B) Symmetric, unimodal. (C) Symmetric, biomodal. (D) Asymmetric, bimodal. (E) Unimodal, skewed to the left. (F) Unimodal, skewed to the right. (G) Rectangular. (H) J-curve.*

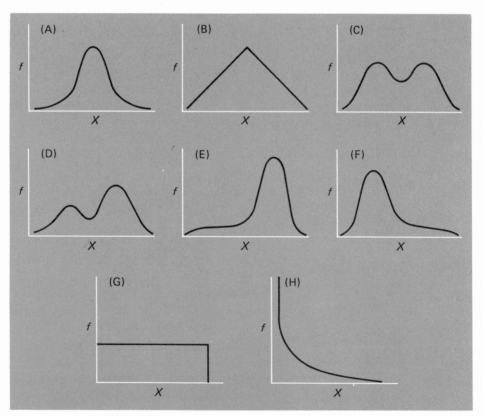

is to the left of the distribution. Such a distribution indicates that most people obtained high scores but some (indicated by the tail) received quite low scores, as might well happen if an inexperienced instructor gave an exam which proved easy for all but the poorest students. Distribution F, on the other hand, is *skewed to the right* (or positively skewed), as is indicated by the position of the tail. This distribution indicates many low scores and some quite high scores, as could easily occur if an equally inexperienced instructor gave a difficult examination on which only the best students performed respectably.

Modality. Modality refers to the number of clearly distinguishable high points or "peaks" in a distribution. In Figure 2.4, Distributions A, B, E, and F are described as *unimodal* because there is one peak; Distributions C and D are *bimodal* because there are two clearly pronounced peaks; and distribution G has no modes at all. Be careful, however, not to be influenced by minor fluctuations when deciding whether to consider a distribution unimodal or bimodal; for example, the curve shown in Figure 2.2 is properly considered as unimodal (with just one less case in the 24–26 interval and one more case in the 27–29 interval, the second "peak" would disappear entirely). A bimodal distribution often indicates that two major kinds of cases are concealed within the one distribution. If you were to plot the height of a group of 1000 adults chosen at random, the resulting distribution would be bimodal, with one peak corresponding to the typical height of women and one peak corresponding to the typical height of men.

Special distributions. Certain frequency distributions have names of their own. A particular bell-shaped, symmetric and unimodal distribution shown in Example A of Figure 2.4 is called the *normal distribution*, about which much more will be said in subsequent chapters. In this distribution, frequencies are greatest at the center, and more extreme scores (both lower and higher) are less frequent. Many variables of interest to the behavioral scientist such as intelligence, are normally distributed. Example G illustrates a *rectangular* distribution, in which each score occurs with the same frequency. A distribution approximately like this would occur if one fair die were rolled a large number of times and the number of 1s, 2s, and so on, were recorded. (This would, however, be a discrete distribution.) Example H is an illustration of a *J-curve*, wherein the score of zero is most frequent and the frequencies decrease as the scores become larger. An example of such a distribution would be the number of industrial accidents per worker over a two-month period, where the majority of workers would have no accidents at all, some would have one accident, a few would have two accidents, and very few unfortunate souls would have a larger number of accidents. The mirror image of this distribution, which is shaped almost exactly like a " J," is also quite naturally called a J-curve, but hardly ever seems to occur in practice.

SUMMARY

In *regular frequency distributions*, all score values are listed in the first column and the frequency corresponding to each score is listed to the right of the score in the second column. *Grouped frequency distributions* sacrifice some information for convenience by combining several score values in a single *class interval* so that fewer intervals and corresponding frequencies need be listed. *Cumulative frequency distributions* are obtained by starting with the frequency corresponding to the lowest score and adding up the frequencies as you go along. *Histograms* and *frequency polygons* are two common methods for graphing frequency distributions; both can be used with any kind of data, but histograms are particularly appropriate for discrete data and frequency polygons are particularly appropriate for continuous data.

3 transformed scores I: percentiles

If you are informed that you have obtained a score of 41 on a 50-point examination, you will undoubtedly require additional information in order to determine just how well you did. You can draw some useful conclusions from the fact that your score represents 82% of the total (for example, it is unlikely that you have failed the examination), but you also need to know how your score compares to the specific group of scores in which it appears—the scores of the other students in the class. If the examination has proved easy for most students and there are many high scores, your score of 41 may represent only average (or even below average) performance. If, on the other hand, the examination was a difficult one for most students, your score may be among the highest (or even the highest).

One way of providing this additional information is to transform the original score (called the *raw score*) into a new score that will show at a glance how well you did in comparison to other students in the class. There are several different kinds of transformed scores ; we will discuss one of them—percentiles—in this chapter, and defer a discussion of others (which depend on material in the following chapters) until Chapter 6.

DEFINITION OF PERCENTILES

A percentile is a single number which gives the *percent of cases in the specific reference group with equal or lower scores*. If, for example, your raw score of 41 corresponds to a percentile score of 85, this means that 85 percent of your class obtained equal or lower scores than you did, while 15 percent of the class received higher scores. If instead your raw score of 41 corresponds to the 55th percentile, this would signify that your score was slightly above average ; 55% of the class received equal or lower scores, while 45% obtained higher scores. A score that would place you at the 5th percentile would be a cause for concern since 95% of the class did better and only 5% did as or more poorly. Thus, the percentile shows directly how an individual score compares to the scores of a specific group.

It is important to keep in mind that you cannot interpret a percentile correctly unless you take careful note of the reference group in question. For example, a college senior who scores at the 90th percentile on an examination would seem to have done well, since his score places him just within the top 10% of some reference group. If, however, this group consists of inmates of an institution for mental defectives, the student should not feel proud of his performance! Similarly, a score at the 12th percentile looks poor, since only 12% of the reference group did as badly or worse; but if the score was obtained by a high-school freshman and the reference group consists of college graduates, the score may actually represent good performance.

It is unlikely that anyone would err in extreme situations such as the foregoing, but there are many practical situations where misleading conclusions can easily be drawn. For example, scoring at the 85th percentile on the Graduate Record Examination, where the reference group consists of college graduates, is superior to scoring at the 85th percentile on a test of general intelligence, where the reference group consists of the whole population (including those people not intellectually capable of obtaining a college degree). If you score at the 60th percentile on the midterm examination in statistics and a friend in a different class scores at the 90th percentile on his statistics midterm, he is not necessarily superior; the students in his class might be poorer, which would make it easier for him to obtain a high standing in comparison to his reference group. Remembering that a percentile *compares* a score to a *specific group of scores* will help you to avoid pitfalls such as these.

COMPUTATIONAL PROCEDURES

Case 1. *Given a raw score, compute the corresponding percentile.*

In order to illustrate the computation of percentiles, let us suppose that you have received a score of 41 in the hypothetical examination data illustrated in Table 2.3 in the preceding chapter. The grouped and cumulative frequency distributions for these data (Table 2.4) are reproduced in Table 3.1.

Table 3.1 *Hypothetical midterm examination scores for 85 students:transforming a raw score of 41 to a percentile rank*

class interval	frequency (f)	cumulative frequency (cf)
48–50	1 ⎫	85
45–47	3 ⎬ 8	84
42–44	4 ⎭	81
39–41	6	77
36–38	7 ⎫	71
33–35	9	64
30–32	14	55
27–29	8	41
24–26	10 ⎬ 71	33
21–23	8	23
18–20	4	15
15–17	3	11
12–14	3	8
9–11	5 ⎭	5

To find the percentile corresponding to the raw score of 41, simply do the following:

1. Locate the class interval in which the raw score falls. (This interval has been boxed in Table 3.1.) Let us call this the "critical interval."

2. Combine the frequencies (f) into three categories: those corresponding to all scores *higher* than the critical interval, those corresponding to all scores in the critical interval, and those corresponding to all scores *lower* than the critical interval, as follows:

	f	percent ($= f/N$)
All *higher* intervals	8	8/85 = 9.4% (H%)
Critical interval (39–41)	6	6/85 = 7.1% (I%)
All *lower* intervals	71	71/85 = 83.5% (L%)
		Check: = 100.0%

As is shown in Table 3.1, a total of 8 people obtained scores higher than the critical interval; 6 people obtained scores in the critical interval; and 71 people obtained scores lower than the critical interval. The last figure is readily obtained by referring to the *cumulative* frequency for the interval just below the critical interval, which shows that 71 people obtained scores of 36–38 or less. Each frequency is then converted to a percent by dividing by N, the total number of people (in this example, 85). We will denote the percent of people scoring in intervals higher than the critical interval by H% (for *higher*), the percent of people scoring in the critical interval by I% (for *in*), and the percent of people scoring lower than the critical interval by L% (for *lower*).

3. It is now apparent that your score of 41 is better than at least 83.5% of the scores, namely those below the critical interval. Thus, your rank in the class expressed as a percentile (or, more simply, your *percentile rank*) must be at least 83.5%. It is also apparent that 9.4% of the scores, the ones above the critical interval, are better than yours. But what of the 7.1% of the scores within the critical interval? It would be too optimistic to assume arbitrarily that your score is higher than the scores of all the other people in the critical interval, since some of these people may well have obtained equal scores of 41. On the other hand, it would be too pessimistic to assume that you did not do better than anyone in your class interval. The solution is to look at your score in comparison to the size of the interval; the higher your score in relation to the critical interval, the more people in that interval you may assume that you outscored.

In order to determine accurately your standing in the critical interval, you must first ascertain the *lower real limit* of the interval. It may seem as though the lower limit of the 39–41 interval is 39, but appearances are often deceiving. If a score of 38.7 were obtained, in which interval would it be placed? Since this score is closer to 39 than to 38, it would be tallied in the 39–41 interval. If a score of 38.4 were obtained, it would be tallied in the 36–38 interval because 38.4 is closer to 38 than to 39. Thus, the *real* dividing line between the critical interval of 39–41 and the next lower interval of 36–38 is not 39, but 38.5; any score between 38.5 and 39.0 belongs in the 39–41 interval, and any score between 38.0 and 38.5 belongs in the 36–38 interval. (For a score of exactly 38.5, it would be necessary to flip a coin or use some other random procedure.) A convenient rule is that the lower real limit of an interval is halfway between the lowest score in that interval (39) and the highest score in the next lower interval (38).

Thus, your score of 41 is 2.5 points (41 − 38.5) up from the lower real limit of the interval. Since the size of the interval is 3, this distance expressed as a fraction is equal to 2.5 points/3 points, or .83 of the interval. Consequently, in addition to the 83.5% of the scores that are clearly below yours, you should credit yourself with .83 of the 7.1% of people in your interval, and your percentile rank is equal to

$$83.5\% + (.83)(7.1\%) = 83.5\% + 5.9\%$$
$$= 89.4\%.$$

This procedure is conveniently summarized by the following formula:

$$percentile\ rank = L\% + \left(\frac{score - LRL}{h} \cdot I\%\right)$$

where

> $L\%$ and $I\%$ are obtained from step 2
> score = raw score in question
> LRL = lower real limit of critical interval
> h = interval size

In our example, this is equal to

$$83.5\% + \left(\frac{41 - 38.5}{3} \cdot 7.1\%\right) = 83.5\% + 5.9\%$$
$$= 89.4\%.$$

Thus, your percentile rank is equal to 89.4%, indicating that approximately 89% of the class received equal or lower scores and only about 11% received higher scores.* This result is depicted in Figure 3.1.

This procedure is also suitable for use with regular frequency distributions, where the interval size (h) equals 1. You will find that the fraction

$$\left(\frac{Score - LRL}{h}\right)$$

always equals $\frac{1}{2}$ for regular frequency distributions.

When you are able to use a regular frequency distribution, you will get somewhat more accurate results, since grouped frequency distributions lose information. For purposes of descriptive statistics, however, the loss of accuracy incurred by grouping will usually not be important, and you can follow the guidelines for forming frequency distributions given in the preceding chapter.

---

* This percentile is computed for a particular *point* on the scale—that is, a score of exactly 41.00. The fact that a score of 41 actually occupies the interval 40.5 to 41.5 is deliberately disregarded in this instance so that we can talk of a percentage above and a percentage at or below the given point, which together sum to 100%. Thus, the point is a "razor-thin" dividing line which divides the group into two parts and not an interval in which scores may fall.

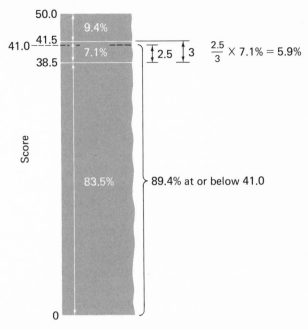

Figure 3.1 *Illustration of percentile rank corresponding to a raw score of 41 for data in Table 3.1.*

Case 2. *Given a percentile, compute the corresponding raw score.*

In the previous problem, you had a raw score (41) and wished to find the corresponding percentile rank. It is also useful to know how to apply the percentile procedures in reverse—to find the raw score that corresponds to a specified percentile value. For example, suppose that an instructor wishes to recommend remedial procedures to the bottom 25% of the class. What raw score should be used as the cutting line? In this example, the percentile (25%) is specified and the raw score is needed, and the steps are as follows:

1. Convert the percentile to a case number by multiplying the percentile by *N*. In the present example, this is equal to $(.25) \times (85)$ or 21.25. Thus, the score that corresponds to the individual whose rank is 21.25 from the bottom of the class (the person scoring at the 25th percentile) is the cutting line that you need.

2. Find the interval in which the case number computed in step 1 falls. This is easily accomplished by starting at the *bottom* of the *cumulative* frequency distribution and proceeding upward until you find the *first* value equal to or greater than the critical case (21.25); the corresponding interval is the "critical interval." This step is illustrated in Table 3.2.

Table 3.2 *Hypothetical midterm examination scores*
for 85 students: finding the raw score corresponding to
a percentile rank of 25%

class interval	frequency (f)	cumulative frequency (cf)	
48–50	1	85	
45–47	3	84	
42–44	4	81	
39–41	6	77	
36–38	7	71	
33–35	9	64	
30–32	14	55	
27–29	8	41	
24–26	10	33	
21–23	8 = f	23	first cf ≥ 21.25
18–20	4	15	
15–17	3	11	
12–14	3	8	15 = SFB
9–11	5	5	

$$pN = .25 \times 85 = 21.25$$

3. The 21.25th case must have a score of at least 20.5, the lower
real limit of the interval in which it appears. However, the
critical interval covers three score points (from 20.5 to 23.5) ;
what point value corresponds to the 25th percentile? The
solution lies in considering how far up from the lower end of
the interval the 21.25th case falls and assigning an appropriate
number of additional score points. If the 21.25th case falls at
the bottom of the interval, very little will be added to 20.5 ;
if the 21.25th case falls near the top of the interval, a larger
quantity will be added to 20.5. In our present example, there
are 15 cases *below* the critical interval, so the 21.25th case
falls (21.25 − 15) or 6.25 cases up the interval. The total
number of cases in the interval is 8, so this distance expressed
as a fraction is 6.25 cases/8 cases or .78. Therefore, in addition
to the lower real limit of 20.5, .78 of the three points included in
the critical interval must be added in order to determine the
point corresponding to the 21.25th case. The desired cutting
score is therefore equal to

$$20.5 + (.78 \times 3) = 20.5 + 2.3$$
$$= 22.8$$

This procedure is conveniently summarized by the following formula:

$$Score_p = LRL + \left(\frac{pN - SFB}{f} \cdot h\right)$$

where

> $Score_p$ = score corresponding to the pth percentile
> LRL = lower real limit of critical interval
> p = specified percentile
> N = total number of cases
> SFB = Sum of Frequencies Below critical interval
> f = frequency within critical interval
> h = interval size

In our example,

$$Score_{.25} = 20.5 + \left(\frac{(.25)(85) - 15}{8} \cdot 3\right)$$

$$= 20.5 + \left(\frac{6.25}{8} \cdot 3\right)$$

$$= 22.8$$

Thus, people with scores of 22 or less are assigned to remedial work, and students with scores of 23 or more are not. The calculations are depicted in Figure 3.2.

Be careful to avoid the temptation to use in the formula any percent that happens to be handy. For example, suppose that the top 10% of the class is to receive a grade of A. What cutting line should be used? Before doing any calculations, you must note that the "top 10%" corresponds to the *90th percentile*; therefore, $p = .90$. The solution:

$$pN = 76.5$$

Critical interval = 39–41

$$Score_{.90} = 38.5 + \frac{76.5 - 71}{6} \cdot 3$$

$$= 38.5 + 2.75$$

$$= 41.25$$

This result should not prove surprising inasmuch as we found in the previous section that a score of 41.00 corresponded to the 89.4th percentile. You will obtain a surprising (and erroneous) result, however, if you unthinkingly set p equal to .10.

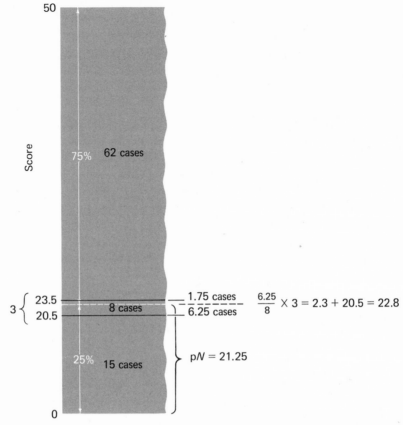

Figure 3.2 *Illustration of raw score corresponding to percentile rank of 25% for data in Table 3.2.*

DECILES, QUARTILES, AND THE MEDIAN

Certain percentile values have specific names, as follows:

percentile		decile	quartile		
90th	=	9th			
80th	=	8th			
75th	=		3rd		
70th	=	7th			
60th	=	6th			
50th	=	5th	=	2nd	=	MEDIAN
40th	=	4th			
30th	=	3rd			
25th	=		1st		
20th	=	2nd			
10th	=	1st			

Thus, whereas the percentile divides the scale into 100 equal parts, the *decile* divides the scale into 10 equal parts and the *quartile* divides the scale into four equal parts. The score corresponding to the 50th percentile has the unique property that exactly half the scores in the group are higher and exactly half the scores in the group are equal or lower. It is called the *median*, and is one of the measures of central tendency that will be discussed in the next chapter.

SUMMARY

The *percentile rank* corresponding to a given score refers to the percent of cases in a given reference group scoring *at or below* that score. A specified raw score may be converted to the corresponding percentile rank to express its standing relative to the reference group, or the raw score corresponding to a specified percentile may be determined.

4 measures of central tendency

The techniques presented in Chapter 2 are useful when it is worth the time and trouble to provide a detailed summary of all of the data in a convenient format. Often, however, your primary objective will be to highlight certain particularly important characteristics of a group of data as a whole. For example, suppose that an inquisitive friend wishes to know how well you are doing in college. You might choose to collect hastily all of your semester grade reports and compile a regular frequency distribution such as the following:

grade	f
A	4
B	9
C	6
D	1
F	0

This would readily indicate that you had received four grades of A, nine grades of B, and so forth. All this detail, however, is not essential in order to answer the question, and it would undoubtedly prove tiresome both for you and for your audience. In addition, presenting the data in this form would make it awkward for your friend to compare your performance to his own college grades. A better plan would be to select one or two important attributes of this set of data and summarize them so that they could be reported quickly and conveniently.

One item of information that you would undoubtedly want to convey would be a number giving the general *location* of the distribution of grades. You could simply state verbally that your college work was slightly below the B level; if you wished to be precise, you would report your numerical grade-point average—a single number that describes the general location of this set of scores. In either case, you would sum up your performance by referring to a central point of the distribution that would be representative; it would clearly be misleading to describe your overall performance as being at the A or D level, even though you did receive some such grades.

This is just one of many situations which benefit from the use of a *measure of "central tendency"—a single number that describes the location of a set of scores.* Other examples would include the average income of men in the United States, the number of cents gained or lost by an average share of stock on the New York Stock Exchange in a single day, and the number of seconds taken by the average rat to run a T maze after 24 hours of food deprivation.

It should be stressed at the outset, however, that *the location of a set of data is not its only important attribute.* Suppose that the average score on a statistics quiz that you have just taken is 5.0. This average provides information as to the general location of the set of quiz scores, but it does not tell you how many high and low scores were obtained. Consequently, you cannot determine what score will be needed to ensure an A (or to just pass with a D!). In a distribution such as the following one, a score of 7 would rank very highly:

7 7 6 5 4 4 4 3

On the other hand, a score of 7 would not seem so illustrious in a distribution like this:

10 10 9 7 5 4 3 2 0 0

In both examples, however, the average, measured as the mean, is equal to 5.0. This indicates that there are important aspects of a set of data which are *not* conveyed by a measure of central tendency; a second vital characteristic will be considered in the next chapter.

THE MEAN

Computation. The *mean* of a set of scores is computed by adding up all the scores and dividing the result by the number of scores. In symbols,

$$\bar{X} = \frac{\sum X}{N}$$

where

$\bar{X}$ = sample mean
$\sum X$ = sum of the X scores (see Chapter 1)
N = total number of scores

Thus, in the case of the first set of quiz scores given above, the mean is equal to

$$\frac{7+7+6+5+4+4+4+3}{8} = \frac{40}{8} = 5.0$$

Note that simply computing $\sum X$ is not sufficient to identify the location of these scores; for example, two scores of 20 also yield a sum of 40. What is further required is to divide by N (the number of scores). This step ensures that the means of two different samples will be comparable, even if they are based on different numbers of scores. For example, the mean of the second set of quiz scores presented in the previous section is equal to 50/10 or 5.0; $\sum X$ is different, but the mean correctly shows that the location is actually the same.

In the case of populations, the Greek letter mu, μ, is used to represent the *population mean*, but the procedure is the same—sum all scores and divide by N.*

Computation from a regular frequency distribution. If scores are available in the form of a regular frequency distribution, the mean is most easily computed from the following formula:

$$\bar{X} = \frac{\sum fX}{N}$$

where $\bar{X}$ = sample mean
  $fX = X$ score multiplied by the frequency of that score
  $\sum fX$ = sum of fX values
  N = total number of scores

Perhaps because of the extra symbol f in the numerator, this equation is often a source of confusion. In fact, it gives exactly the same result as would the preceding formula applied to the same data in an untabulated format. It may be helpful to think of the first formula for the sample mean as a special case of the second in which $f = 1$. The two procedures are compared in Figure 4.1; note that the value $N = 20$ is easily recovered from the regular frequency distribution by computing $\sum f$ (the total of the *frequencies* shows *how many scores* there are).

As was mentioned in Chapter 2, means (and other statistics) should in general not be computed from *grouped* frequency distributions (which do not give the exact value of every score), although

---

* The distinction between samples and populations was introduced in Chapter 1, and it will be treated at greater length in Chapter 7.

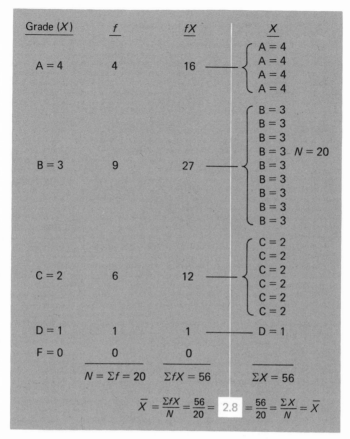

Grade (X)	f	fX	X
A = 4	4	16	A = 4 A = 4 A = 4 A = 4
B = 3	9	27	B = 3 B = 3 B = 3 B = 3 N = 20 B = 3 B = 3 B = 3 B = 3 B = 3
C = 2	6	12	C = 2 C = 2 C = 2 C = 2 C = 2 C = 2
D = 1	1	1	D = 1
F = 0	0	0	
	$N = \Sigma f = 20$	$\Sigma fX = 56$	$\Sigma X = 56$

$$\bar{X} = \frac{\Sigma fX}{N} = \frac{56}{20} = 2.8 = \frac{56}{20} = \frac{\Sigma X}{N} = \bar{X}$$

Figure 4.1   *The mean computed from a regular frequency distribution and from untabulated data.*

they may be approximated by treating all the scores in any interval as if they fell at the midpoint of the interval.

Interpretation. One important characteristic of the mean is that the sum of distances (or *deviations*) of all scores from the mean is zero. That is,

$$\sum (X - \bar{X}) = 0$$

It can readily be proved that this must always be true.* As an

$$*\sum (X - \bar{X}) = \sum X - N\bar{X} \text{ (Rule 7, Chapter 1)}$$

but

$$\sum X = N\bar{X}$$

(because $\bar{X} = \left(\sum X\right)/N$). Therefore,

$$\sum (X - \bar{X}) = N\bar{X} - N\bar{X}$$
$$= 0$$

illustration, consider once again the small set of quiz scores discussed previously:

score	deviation from mean $(X - \bar{X})$
7	+2
7	+2
6	+1
5	0 $(\bar{X} = 5)$
4	−1
4	−1
4	−1
3	−2

$$\sum (X - \bar{X}) = +5 - 5 = 0$$

Thus, the mean balances or equates the sums of the positive and negative deviations. It is in this sense that it gives the location of the distribution. This implies, however, that the mean will be sensitive to extreme values on one side that are not balanced by extreme values on the other side, as the following example shows:

score	$X - \bar{X}$	
39	+30	$(\bar{X} = 9)$
7	−2	
6	−3	
5	−4	
4	−5	
4	−5	
4	−5	
3	−6	

$$\sum (X - \bar{X}) = +30 - 30 = 0$$

As a result of changing one score from 7 to 39, the mean shows a substantial four-point increase. In addition, all of the other seven scores now fall below the mean in order to balance the effects of the large positive deviation introduced by the score of 39. One might well question the use of the mean to describe the location of a set of data in a situation where it is so influenced by one extreme score!

As another example of the sensitivity of the mean to extreme values, consider the case of an unethical manufacturing company in which the president earns one million dollars per year and the 99 assembly line workers earn only $2000 per year. The president might attempt to refute criticism of the company's miserly tactics by arguing that the mean annual income of these 100 people is $11,980; the workers would undoubtedly object to the appropriateness of this figure. Here again, the use of the mean as the index of location is indeed questionable. It almost always is with income data, since almost all income distributions are positively skewed.

Usage. The mean has many advantageous properties. First, it takes all of the scores into account, and therefore makes the most of the information provided by the data. Second, it is the most stable of the measures of central tendency for most distributions encountered in practice; it is the most consistent across different samples drawn from the same population. For this and other reasons, many of the procedures of inferential statistics make use of the mean. Thus, a third advantage of the mean is that it is usable as a datum in further statistical analyses, while other measures of central tendency usually are not. For these reasons, the mean is the most frequently suitable measure of central tendency.

At times, however, the first advantage becomes a liability instead of an asset. As we have seen, extreme scores at one end of a distribution exert a strong influence on the mean and cause it to give a misleading picture of the location of the distribution. Therefore, when a distribution is highly skewed and when you do not intend to use the measure of central tendency in subsequent statistical analyses, you should seek an alternative to the mean which will not be affected by unbalanced extreme scores.

In some cases, the actual size of the extreme scores may be unknown. For example, suppose that a few subjects in a learning experiment do not learn the task even after a great many trials. Rather than wait perhaps hours in the hope that learning would occur, you might terminate the experiment for anyone failing to learn after (say) 75 trials. Such subjects would therefore have learning scores of "at least 75 trials." They would therefore be the slowest learners (and should not be discarded from the experiment, as the results would then be biased); but their exact scores would be unknown, so you could not compute a sample mean.

In situations such as these, a measure of central tendency is needed that does *not* take the exact value of extreme scores into account. Such a measure is available, and it is called the median.

THE MEDIAN

Computation. The median (Mdn) is defined as the score corresponding to the 50th percentile. It is computed using the procedures given in the previous chapter (simply compute *Score* $_{.50}$).

Interpretation. The median is the *middle* score in the distribution when scores are put in order of size. It is not affected by the size of extreme values; the median income in the unethical manufacturing company described in the preceding section is $2000.01 (using an interval size of one dollar). The median would be the same even if the company president took home five million dollars per year instead of one million. Similarly, in the learning experiment, you would simply call the highest class interval "75 trials or more" and compute the median, which is not affected by the numerical value of extremely high (or low) scores.

As the example of the manufacturing company indicates, the median ($2000.01) will be smaller than the mean ($11,980) in a *positively* skewed distribution (where the extreme values are at the higher end), because the mean will be pulled upward by the extreme high scores. In a *negatively* skewed distribution, where the extreme values are at the lower end, the median will be larger than the mean (which will be pulled downward by the extreme low scores). In a symmetric distribution, where there is no skew, the mean and median will be equal.

Usage. You should use the median when you do not intend to use the measure of location in subsequent statistical analyses (that is, your objectives are purely descriptive) *and* when either (1) the data are highly skewed or (2) there are inexact data at the extremes of the distribution. This will enable you to profit from the fact that the median is not affected by the size of extreme values and obtain the most reasonable description of the location of such distributions. In almost all other cases, use the mean so as to benefit from its numerous advantages.

THE MODE

The mode is simply the score that occurs most often. For example, the mode of the data in Figure 4.1 is 3 (or B) since this score was obtained more often than any of the others (9 times). The mode is a crude descriptive measure of location which ignores a substantial part of the data; therefore, it is not usually used in research in the behavioral sciences. If you were to record the results of a roulette wheel on a few thousand trials, however, your best bet on subsequent spins would be the mode, in the hope that you had identified the single case most likely to occur in one spin of a flawed wheel. (If the wheel is unbiased, the mode of the previous trials is as good—or as bad—a choice as any other.)

SUMMARY

One important attribute of a set of scores is its *location,* that is, where in the possible range between minus infinity and plus infinity the scores tend to fall. This can be described in a single number by using either the *mean,* the best measure in most instances ; or the *median* (the score corresponding to the 50th percentile), preferable when data are highly skewed or there are extreme data whose exact values are unknown, and when the objectives are purely descriptive.

5 measures of variability

In addition to general location, there is a second important attribute of a distribution of scores—its *variability*. Measures of variability are used extensively in the behavioral sciences, and it is therefore essential to understand thoroughly the meaning of this concept as well as the calculational procedures.

THE CONCEPT OF VARIABILITY

Variability refers to how *spread out or scattered* the scores in a distribution are (or, how like or unlike each other they are). As an illustration, a few distributions involving a small number of scores are shown below.

Distribution 1 :	7.0	7.0	7.0	7.0	7.0	7.0			
Distribution 2 :	7.0	7.1	7.1	7.1	7.1	7.2			
Distribution 3 :	40.0	40.1	40.1	40.1	40.1	40.2			
Distribution 4 :	7.0	7.0	6.0	5.0	4.0	4.0	4.0	3.0	
Distribution 5 :	10.0	10.0	9.0	7.0	5.0	4.0	3.0	2.0	0.0 0.0
Distribution 6 :	97.8	88.5	83.4	76.2	69.9	67.3	58.4	44.7	

The minimum possible variability is zero, and this will occur only if all the scores are exactly the same (as in Distribution 1) and there is no variation at all. In Distribution 2, there is a very small amount of variability; the scores are somewhat spread out, but only to a very slight extent. Distribution 3 is equal in variability to Distribution 2. The locations of these two distributions differ, but variability is not dependent on location; the distance between each score and any other (and hence the amount of spread) is identical. Each of the remaining three distributions is more variable than the ones that precede it. At the opposite extreme to Distribution 1, one might postulate a distribution with scores spread out over the entire range from −1,000,000 to +1,000,000 (or more). Such extreme variabilities, however, are rarely encountered in practice.

Variability is important in many areas of endeavor, although it is frequently not reported (or described vaguely in words) because it is less familiar to nontechnical audiences than is central tendency. We list a few examples:

Testing. Suppose that you get a score of 75 on a statistics midterm examination and that the mean of the class is 65; the maximum possible score is 100. Although your score cannot be poor because it is above average, its worth in comparison to the rest of the class will be strongly influenced by the variability of the distribution of examination scores. If most scores are clustered tightly around the mean of 65, your score of ten points above average will stand out as one of the highest (and may well merit a grade of A). If, on the other hand, the scores are widely scattered and values in the 80s and 90s (and 40s and 50s) are frequently obtained, being ten points above average will not be exceptional because many will have done better, and in this case your score may be worth no more than a B— or C+. Thus, a mean of 65 together with low variability would indicate that you did very well; a mean of 65 together with high variability would imply that your performance was less than outstanding compared to the group that took the test. These two possibilities are illustrated graphically in Figure 5.1; note that the spread of a frequency polygon indicates the variability of the distribution.

As a similar example, consider Distributions 4 and 5, which were discussed in Chapter 4. The mean of both distributions is 5.0. However, a score of 7 ranks higher in Distribution 4, where the

Figure 5.1 *Frequency polygons of two distributions with the same mean but different variability.*

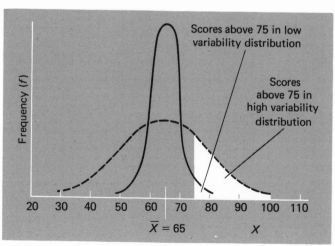

scores are less variable, than in Distribution 5, where the scores are more variable. Looking at the lower end of these distributions, a score of 3 is poorer in comparison to the group in Distribution 4. Thus, being two points above or below average stands out more in a distribution in which there is less spread.

The stock market. Evening television news programs usually report the average gain or loss of stocks on the various markets at the end of a day of trading. For example, on a given day an average share on the New York Stock Exchange may lose five cents. This value of −5¢ (where the minus sign indicates a loss) is a mean, computed by adding up the gains and losses recorded by each stock on the Exchange (scoring losses as negative numbers) and dividing by the total number of different stocks. It would also be useful, however, to know the variability of the distribution of gains and losses. If the variability is small, most stocks were close to the mean value of a five-cent loss; there were few big gains or big losses. This additional information would tell the jittery stockholder that while each of his stocks probably incurred a minute loss, none was likely to have suffered a large loss (or registered a sizable gain). On the other hand, suppose that the variability of the distribution of gains and losses is large. This would indicate that there was wide variation from the mean of −5¢, with some stocks losing much more and some stocks gaining appreciably. In this case, the stockholder would want to rush to his newspaper and check each of his holdings in the listing of individual stocks, because he could not rely as surely on the average as indicative of the performance of any one stock.

Sports. Suppose that two professional basketball players each average 20.3 points per game. Although their mean performance is the same, they may be quite different in other respects. Suppose that one is consistent (low variability) and always scores right around 20 points in each game, rarely hitting as high as 25 but also rarely falling to 15 or less. The second player, on the other hand, is erratic (high variability) ; he scores over 30 points in some games but also drops to below 10 in others. If nothing else, the second player is likely to be a much greater source of frustration to his coach and to the fans. Here again, a measure of variability would provide useful additional information to that given by the mean.

Psychology. Modern psychology is predicated upon the idea that people differ. On any trait of interest—intelligence, throwing a baseball, introversion, height, mathematical ability—people are distributed over the entire range from low to high. If people were all the same, the behavior of the entire population could be predicted from a knowledge of one individual, and psychological research would be virtually unnecessary; anything true about you would be true about everyone else. Since this is not the case, the psychologist has the essential task of measuring and explaining variation—why some people are more neurotic and others less so, why some people do better in school or on the job than others, why one person performs differently on different occasions, and so on. Thus variability is the *raison d'etre* of the psychologist.

Statistical inference. Yet another reason for the importance of variability will become evident when statistical inference is discussed (Chapters 7ff.); other things being equal, the more variable a phenomenon is, the less precise is the estimate one can get of the population's location (for example, mean) from sample information.

Rather than describe variability in ambiguous terms such as "small" or "large," it is preferable to summarize the variability of a distribution of scores in a single number. Techniques for accomplishing this are discussed below.

THE RANGE

One possible way of summarizing the variability of a distribution is to look at the distance between the smallest and largest scores, and the *range* of a distribution is simply defined as the largest score minus the smallest score. For example, the range of Distribution 4 is $(7.0 - 3.0)$ or 4.0. While this procedure makes intuitive sense, it is likely to give misleading results on many occasions because the extreme values are frequently eccentric and atypical of the rest of the distribution. Consider the following two distributions:

Distribution A: 10 10 10 9 7 6 5 4 4 3 2 0 0
Distribution B: 10 6 6 5 5 5 5 5 5 4 4 0

In both of these examples, the range is equal to 10. However, Distribution A, with scores spread out over the entire 10-point range, is clearly more variable than Distribution B, where all but two scores are concentrated near the middle of the distribution. The range is a poor measure of the variability of Distribution B because the extreme values are not typical of the total variation in the distribution; if the two extreme scores are excluded, the range drops to 2. Since this type of distortion occurs fairly often, the range (like the mode) is best regarded as a crude measure that should not generally be used in behavioral science research.

THE STANDARD DEVIATION AND VARIANCE

We have seen that the mean, an average which takes all of the scores into account, is usually the best measure of central tendency. Similarly, an "average" variability that is based on all the scores will usually provide the most accurate information. Before an *average* variability can be computed, however, this concept must be defined in terms of an individual score. Variability actually refers to the difference between each score and every other score, but it would be quite tedious to compute this in practice (especially if N is large). For example, if there are 100 scores, you would have to compute the difference between the first score and each of the 99 other scores, and then compute the difference between the second score and each of the 98 remaining scores, and so on.

A more feasible plan, which will serve the purpose equally well, is to define the "differentness" or *deviation* of a single score in terms of how far it is from the *average*, defined by one of the measures described in Chapter 4, of the scores. This will ultimately lead to a useful measure of variability because the average is a summary of the collective location of the scores. Thus, a distribution of scores that is closely packed together will have most scores close to each other, and hence close to their average, while a highly variable distribution will have many scores quite a distance from each other, and hence far from their average. Since the mean is the most frequently used measure of central tendency, a reasonable procedure is to define the deviation of a single score as its *distance from the mean**:

$$\text{Deviation score} = X - \bar{X}$$

Lower case letters are sometimes used to represent deviation scores; thus, $x = X - \bar{X}$. Extremely deviant scores (ones far away from the mean) will have numerically large deviation scores, while scores close to the mean will have numerically small deviation scores.

The next step is to derive a measure of variability that will take into account the deviations of all the scores. There are several

---

* A reference point other than the mean could be used, but the choice of the mean has certain statistical advantages as well as making good intuitive sense. For example, it can be proved that the mean is the value of c about which $\sum (X - c)^2$, the sum of squared deviations, is a minimum (the importance of which will become apparent in the following discussion). The mean, however, is *not* the point about which the sum of deviations whose signs are ignored is a minimum; the *median* is.

possible ways to do this. If we were to average the deviation scores by the usual procedure of summing and dividing by N, we would get

$$\frac{\sum (X - \bar{X})}{N}$$

It will prove extremely frustrating to try and use this as the measure of variability because (as was proved in Chapter 4) $\sum (X - \bar{X})$ is *always* equal to zero. As a result, this "measure" cannot possibly provide any information as to the variability of any distribution.

This problem could be overcome if all of the deviations were positive. One possibility, therefore, would be first to take the *absolute value* of each deviation, its numerical value ignoring the sign, and then compute the average variability. That is, we could compute

$$\frac{\sum |X - \bar{X}|}{N}$$

where $|X - \bar{X}|$ is the absolute value of the deviation from the mean. This is not an unreasonable procedure (and in fact yields a descriptive measure called the *average deviation*), but is usually rejected because absolute values are unsuitable for use in further statistical analysis. The measure which is most frequently used circumvents this difficulty by *squaring* each of the deviations prior to taking the average. The measure of variability thus produced is called the *variance*, symbolized by σ^2:

$$\sigma^2 = \frac{\sum (X - \bar{X})^2}{N}$$

This is a basic measure of the variability of any set of data. However, when the data of a sample are to be used to estimate the variance of the population from which the sample was drawn, the *population variance estimate* (symbolized by s^2) is computed instead:

$$s^2 = \frac{\sum (X - \bar{X})^2}{N - 1}$$

In each of the above formulas, the order of operations is: (1) subtract the mean from each score; (2) square each result; (3) sum; (4) divide. The sample estimate of the population variance, s^2, is computed a bit differently from σ^2; the sum of squared deviations is divided by $N-1$ instead of N. This is to enable s^2 to be an *unbiased* estimate of the population variance—that is, an estimate that on the average will be too large as often as it is too small.

There is one remaining difficulty. Having squared the deviations to eliminate the negative numbers that otherwise would have led to a total of zero with annoying regularity, the variance is in terms of original units *squared*. For example, if you are measuring IQ, the variance indexes variability in terms of *squared IQ deviations*. It is frequently preferable to have a measure of variability that is in the same units as the original measure, and this can be accomplished by taking the positive square root of the variance. This yields a commonly used measure of variability called the *standard deviation*, symbolized by σ or s depending on whether the variance or the population variance estimate is used:

$$\sigma = +\sqrt{\sigma^2} = \sqrt{\frac{\sum(X - \bar{X})^2}{N}}$$

$$s = +\sqrt{s^2} = \sqrt{\frac{\sum(X - \bar{X})^2}{N-1}}$$

Whereas the mean represents the "average *score*," the standard deviation represents a kind of "average *variability*"—the average of the deviations of each score from the mean $(X - \bar{X})$—subject only to two minor complications: squaring and subsequently taking the positive square root, to eliminate the minus signs before averaging and return to the original unit of measurement afterwards; and dividing by $N-1$ instead of N when estimates of the population are involved. The formulas given above are called the *definition* formulas for σ and s because their primary function is to define the meaning of these terms; they are not necessarily the formulas by which σ and s are most easily computed.

Illustrative examples of the computation of σ^2 and σ, and s^2 and s using the definition formulas are shown on the left-hand side of Table 5.1. Note that a partial check on the calculations is possible in that $\sum(X - \bar{X})$ should always equal zero. As expected from the previous discussion concerning these distributions, Distribution 5 has a larger standard deviation (that is, is more variable) than Distribution 4; the "average" deviation from the mean is about four points in Distribution 5, but is only 1.5 points in Distribution 4. It does not matter that no score is exactly 1.5 points above or below the mean of Distribution 4. The sample mean often turns out to be a number that did not occur (or, in some cases, could not possibly have occurred) in the sample,* yet it is nevertheless a good summary of the location of the distribution. Similar reasoning applies in the case of variability and the standard deviation.

Computing formulas. Using the definition formulas to calculate σ and s can be awkward for several reasons. If the mean is not a whole number, subtracting it from each score will yield a deviation

Table 5.1 *Computation of σ^2 and σ, and s^2 and s, for two small samples using the definition and computing formulas*

Example 1. Distribution 4 ($\bar{X} = 5.0$)

X	$X - \bar{X}$	$(X - \bar{X})^2$	X	X^2
7	$7 - 5 =$ 2	4	7	49
7	$7 - 5 =$ 2	4	7	49
6	$6 - 5 =$ 1	1	6	36
5	$5 - 5 =$ 0	0	5	25
4	$4 - 5 = -1$	1	4	16
4	$4 - 5 = -1$	1	4	16
4	$4 - 5 = -1$	1	4	16
3	$3 - 5 = -2$	4	3	9

$$\sum (X - \bar{X})^2 = 16 \qquad \sum X = 40 \qquad \sum X^2 = 216$$

$$\sigma^2 = \frac{\sum (X - \bar{X})^2}{N} = \frac{16}{8} = 2.00 \qquad \sigma^2 = \frac{1}{N}\left[\sum X^2 - \frac{(\sum X)^2}{N}\right]$$

$$\sigma = \sqrt{2.00} = 1.41 \qquad\qquad = \frac{1}{8}\left[216 - \frac{(40)^2}{8}\right]$$

$$= \frac{1}{8}(216 - 200)$$

$$= \frac{1}{8}(16) = 2.00$$

$$\sigma = \sqrt{2.00} = 1.41$$

$$s^2 = \frac{\sum (X - \bar{X})^2}{N - 1} = \frac{16}{7} = 2.29 \qquad s^2 = \frac{1}{N - 1}\left[\sum X^2 - \frac{(\sum X)^2}{N}\right]$$

$$s = \sqrt{2.29} = 1.51 \qquad\qquad = \frac{1}{7}\left[216 - \frac{(40)^2}{8}\right]$$

$$= \frac{1}{7}(216 - 200) = \frac{1}{7}(16) = 2.29$$

$$s = \sqrt{2.29} = 1.51$$

* For example, the mean of an examination on which no fractional scores were given might be 67.3.

Table 5.1   *(Continued)*

Example 2. Distribution 5 ($\bar{X} = 5.0$)

X	$X - \bar{X}$	$(X - \bar{X})^2$	X	X^2
10	$10 - 5 =$ 5	25	10	100
10	$10 - 5 =$ 5	25	10	100
9	$9 - 5 =$ 4	16	9	81
7	$7 - 5 =$ 2	4	7	49
5	$5 - 5 =$ 0	0	5	25
4	$4 - 5 = -1$	1	4	16
3	$3 - 5 = -2$	4	3	9
2	$2 - 5 = -3$	9	2	4
0	$0 - 5 = -5$	25	0	0
0	$0 - 5 = -5$	25	0	0

$$\sum (X - \bar{X})^2 = 134 \qquad \sum X = 50 \qquad \sum X^2 = 384$$

$$\sigma^2 = \frac{\sum (X - \bar{X})^2}{N} = \frac{134}{10} = 13.40$$

$$\sigma = \sqrt{13.40} = 3.66$$

$$\sigma^2 = \frac{1}{N} \left[\sum X^2 - \frac{(\sum X)^2}{N} \right]$$

$$= \frac{1}{10} \left[384 - \frac{(50)^2}{10} \right]$$

$$= \frac{1}{10} (384 - 250)$$

$$= \frac{1}{10} (134) = 13.40$$

$$\sigma = \sqrt{13.40} = 3.66$$

$$s^2 = \frac{\sum (X - \bar{X})^2}{N - 1} = \frac{134}{9} = 14.89$$

$$s = \sqrt{14.89} = 3.86$$

$$s^2 = \frac{1}{N - 1} \left[\sum X^2 - \frac{(\sum X)^2}{N} \right]$$

$$= \frac{1}{9} \left[384 - \frac{(50)^2}{10} \right]$$

$$= \frac{1}{9} (384 - 250) = \frac{1}{9} (134) = 14.89$$

$$s = \sqrt{14.89} = 3.86$$

score with decimal places, which when squared will produce still more decimal places. This can make the computations quite tedious. Second, the mean must be calculated before the formula can be used, which necessitates two steps in the computation of the standard deviation (although normally the mean will be desired anyway). Therefore, *computing formulas* for σ and s have been derived by manipulating the definition formulas algebraically. The computing formulas and definition formulas yield identical results, but the computing formulas are designed to facilitate hand or desk-calculator computation. The computing formulas for the standard deviation and variance are:

$$\sigma^2 = \frac{1}{N}\left[\sum X^2 - \frac{(\sum X)^2}{N}\right], \qquad \sigma = \sqrt{\frac{1}{N}\left[\sum X^2 - \frac{(\sum X)^2}{N}\right]}$$

$$s^2 = \frac{1}{N-1}\left[\sum X^2 - \frac{(\sum X)^2}{N}\right], \qquad s = \sqrt{\frac{1}{N-1}\left[\sum X^2 - \frac{(\sum X)^2}{N}\right]}$$

Examples of the use of the computing formula for σ^2 and σ, and s^2 and s, are shown in Table 5.1.

Computing the standard deviation and variance from a regular frequency distribution. When scores are arranged in the form of a regular frequency distribution, it is convenient to determine the squared deviation (or X^2 in the computing formula) just once for each score value and then multiply it by the frequency of that score value. That is:

definition formulas   *computing formulas*

$$\sigma^2 = \frac{\sum f(X - \bar{X})^2}{N} \qquad \sigma^2 = \frac{1}{N}\left[\sum fX^2 - \frac{(\sum fX)^2}{N}\right]$$

$$\sigma = \sqrt{\frac{\sum f(X - \bar{X})^2}{N}} \qquad \sigma = \sqrt{\frac{1}{N}\left[\sum fX^2 - \frac{(\sum fX)^2}{N}\right]}$$

$$s^2 = \frac{\sum f(X - \bar{X})^2}{N-1} \qquad s^2 = \frac{1}{N-1}\left[\sum fX^2 - \frac{(\sum fX)^2}{N}\right]$$

$$s = \sqrt{\frac{\sum f(X - \bar{X})^2}{N-1}} \qquad s = \sqrt{\frac{1}{N-1}\left[\sum fX^2 - \frac{(\sum fX)^2}{N}\right]}$$

Despite the presence of an extra f at various places, these formulas when applied to a regular frequency distribution yield identical results to the formulas given previously when applied to untabulated data. As an illustration, Distribution 4 has been arranged into a regular frequency distribution, and s^2 and s have been computed in Table 5.2. Compare the computations to those for Distribution 4 in Table 5.1 and verify that the procedures are in effect identical. You may also wish to use these data to verify that the corresponding formulas for σ^2 and σ are equivalent.

Table 5.2 *Computation of s^2 and s from a regular frequency distribution using the definition and computing formulas*

Distribution 4 ($\bar{X} = 5.0$)

X	f	$X-\bar{X}$	$(X-\bar{X})^2$	$f(X-\bar{X})^2$	X	f	X^2	fX	fX^2
7	2	$7-5=$ 2	4	8	7	2	49	14	98
6	1	$6-5=$ 1	1	1	6	1	36	6	36
5	1	$5-5=$ 0	0	0	5	1	25	5	25
4	3	$4-5=-1$	1	3	4	3	16	12	48
3	1	$3-5=-2$	4	4	3	1	9	3	9
2	0	$2-5=-3$	9	0	2	0	4	0	0
1	0	$1-5=-4$	16	0	1	0	1	0	0
0	0	$0-5=-5$	25	0	0	0	0	0	0

$$\sum f(X-\bar{X})^2 = 16 \qquad \sum fX = 40 \quad \sum fX^2 = 216$$

$$s^2 = \frac{\sum f(X-\bar{X})^2}{N-1} = \frac{16}{7} = 2.29$$

$$s = \sqrt{2.29} = 1.51$$

$$s^2 = \frac{1}{N-1}\left[\sum fX^2 - \frac{(\sum fX)^2}{N}\right]$$

$$= \frac{1}{7}\left[216 - \frac{(40)^2}{8}\right]$$

$$= \frac{1}{7}(216 - 200) = \frac{1}{7}(16) = 2.29$$

$$s = \sqrt{2.29} = 1.51$$

SUMMARY

A second important attribute of a set of scores is its *variability,* or how much the scores *differ from one another.* In the behavioral sciences, this is customarily summarized in a single number by computing the *variance* or *standard deviation* ; the larger the variance or standard deviation, the more different the numbers are from one another. The concept of variability has many practical applications and is particularly important in statistical work.

6 transformed scores II: Z and T scores

In Chapter 3, we saw that it is often desirable to transform a raw score into a new score (such as a percentile rank) that will show at a glance how the score stands in comparison to a specific reference group. In the preceding two chapters, we have seen that the important characteristics of a group of scores are its location, frequently summarized by $\bar{X}$, and its variability, for which we will use σ. Thus, it is possible to deduce how well a given score compares to the reference group by using the mean and standard deviation of the group, and it is often desirable to build this information right into that score itself—that is, to derive a transformed score that shows at a glance the relationship of the original raw score to the mean, using the standard deviation of the reference group as the unit of measurement.

For example, suppose that a college student takes three midterm examinations in three different subjects and obtains the following *raw* scores:

	English	mathematics	psychology
X	80	65	75

On the surface, it might seem as though the student's best score is in English and his poorest score is in mathematics. It would be highly unwise, however, to jump to such a conclusion, since there are several reasons why the raw scores may not be directly comparable. For example, the English examination may have been easy and resulted in many high scores, while the mathematics examination may have been extremely difficult. Or, the English examination may have been based on a total of 100 points and the mathematics examination on a total of only 80 points. The raw scores do provide information about the absolute number of points earned, but they give no indication as to how good the performance is, and certainly no indication of how good the performance is compared to others.

Now suppose that we specify in addition the mean and standard deviation of each test:

	English	mathematics	psychology
X	80	65	75
$\bar{X}$	85	55	60
σ	10	5	15

This additional information changes the picture considerably. Looking at the means, we can see that the scores on the English examination were high, so much so that the score of 80 is below average. On the other hand, both the mathematics and psychology scores are above average. Therefore, the student's poorest result is in English.

The unwary observer might now conclude that the student's best score is in psychology, since that score is 15 points above average while the mathematics score is only 10 points above average. However, as was pointed out in Chapter 5, the variability of a distribution of scores must also be considered when interpreting the relative standing of a given score. The standard deviation indicates that the " average " variability on the psychology test was 15 points from the mean; some scores were more than 15 units from the mean and some were less. Therefore, the student's psychology score of 75, which is 15 points or one standard deviation above average, was exceeded by a fair number of better scores.* The average variability on the mathematics examination, however, was only five points from the mean. Thus, the student's mathematics score of 65 is 10 points or _two_ standard deviations above average; it is unusually far above the mean and is therefore likely to be one of the best scores.†

Thus, it turns out that the picture presented by the raw scores was quite misleading in this instance. Based on a comparison to each of the groups that took each test, the student's best score is in mathematics, the next best score is in psychology, and the poorest score is in English, all relative to the other students in each course.

The raw scores of 80, 65, and 75 cannot be compared directly because they come from distributions with different means and different standard deviations; thus, the units in which they measure are not the same from test to test. This difficulty can be overcome,

---

* The exact number depends on the shape of the distribution of scores. For example, if the scores are normally distributed, approximately 16% of the scores are more than one standard deviation above the mean (as will be shown in Chapter 8).

† If the distribution is normal, less than $2\frac{1}{2}$% of the scores are more than two standard deviations above the mean.

however, by transforming the scores on each test to a common scale with a specified mean and standard deviation. This new scale would then serve as a "common denominator" which would enable the transformed scores of different tests to be compared directly. Two questions remain: How does one go about changing the scores so that they will have the desired common mean and standard deviation? And, what values of the mean and standard deviation are useful choices for the "common denominator"?

RULES FOR CHANGING $\bar{X}$ AND σ

Suppose that the mathematics instructor in the previous example suffers a pang of conscience over the low mean and decides to add five points to everyone's score. Since he has added a *constant* amount to each and every score (one that is exactly the same for all the scores), he does not have to recompute a new mean via the usual formula; it can readily be proved that adding five points to everyone's score increases the mean by five points.* Thus, the mean of the transformed scores (symbolized by $\bar{X}_{new}$) will be 60. Similarly, it can be proved that:

If a constant k is *subtracted* from every score,

$$\bar{X}_{new} = \bar{X}_{old} - k.$$

If every score is *multiplied* by a constant k,

$$\bar{X}_{new} = k\bar{X}_{old}.$$

If every score is *divided* by a constant k,

$$\bar{X}_{new} = \bar{X}_{old}/k.$$

For example, if the English instructor subtracts 7.5 points from every score, the new mean is 77.5; if he multiplies every score by 4, the new mean is 340; and if he divides every score by 2, the new mean is 42.5. Be sure to note that these rules for conveniently determining the new mean work only if every original score is altered by exactly the same amount.

Insofar as the variability of the distribution is concerned, it is readily proved that adding a constant to every score or subtracting a constant from every score does *not* change the standard deviation or variance. This is intuitively obvious; adding or subtracting a constant does not change the spread of the distribution since

---

* The proofs mentioned in this section are presented in the Appendix at the end of this chapter.

each score is increased or decreased by the same amount. Insofar as multiplication and division by a constant are concerned, it can be shown that:

If every score is *multiplied* by a positive constant k,

$$\sigma_{new} = k\sigma_{old}$$
$$\sigma_{new}^2 = k^2\sigma_{old}^2$$

If every score is *divided* by a positive constant k,

$$\sigma_{new} = \sigma_{old}/k$$
$$\sigma_{new}^2 = \sigma_{old}^2/k^2$$

For example, if the English instructor adds 10 points or subtracts 6 points from every score, the standard deviation remains 10 (and the variance remains 10^2 or 100). If he multiplies every score by 4, the new standard deviation equals 4×10 or 40 and the new variance equals $4^2 \times 100$ or 1600. If he divides every score by 2, the new standard deviation equals 10/2 or 5 and the new variance equals $100/2^2$ or 25.

These rules make it possible to obtain transformed scores with any desired mean and standard deviation. For example, scores on the English test can be transformed to scores comparable to those on the mathematics test in two steps:

procedure	new $\bar{X}$	new σ
1. Divide every score on the English examination by 2.0.	$85/2 = 42.5$	$10/2 = 5$
2. Add 12.5 to each of the scores obtained in Step 1.	$42.5 + 12.5 = 55$	5 (no change)

Note that the first step is to change the standard deviation to the desired value by multiplying or dividing each score by the appropriate constant, which affects the value of both $\bar{X}$ and σ. Then, the desired mean is obtained by adding or subtracting the appropriate constant, which does not cause any further change in σ. When the student's English score of 80 is subjected to these transformations, it becomes $(80/2) + 12.5$ or 52.5, and it is evident that this score is not nearly as good as the student's mathematics score. Note that both the original and the transformed English scores are half a standard deviation below the means of their respective distributions.

STANDARD SCORES (Z SCORES)

Since the techniques discussed in the preceding section make it possible to switch to any new mean and standard deviation, the next logical step is to choose values of $\bar{X}_{new}$ and σ_{new} that facilitate comparisons between scores. It was reasonable to transform the English scores to scores with a new mean of 55 and a new standard deviation of 5 because these values corresponded to the ones on the mathematics test to which we wished to make the comparison. These values would not, however, be good ones to select for a common scale because the transformed score of 52.5 would still not indicate at a glance how good the score is in comparison to the group; you would have to keep in mind that it was based on a mean of 55 and a standard deviation of 5.

A more useful procedure is to convert the original scores to new scores with a mean of 0 and a standard deviation of 1, called *Z* scores or *standard scores*. Standard scores have two major advantages. Since the mean is zero, you can tell at a glance whether a given score is above or below average; an above-average score is positive and a below-average score is negative. Also, since the standard deviation is 1, the numerical size of a standard score indicates *how many standard deviations* above or below average the score is. We saw at the beginning of this chapter that this information offers a valuable clue as to how good the score is; a score one standard deviation above average (that is, a standard score of $+1$) would demarcate approximately the top 16% in a normal distribution, while a score two standard deviations above average (a standard score of $+2$) would demarcate approximately the top $2\frac{1}{2}$% in a normal distribution.

To convert a set of scores to standard scores, the first step is to subtract the original mean from every score. According to the rules given in the previous section, the new mean is equal to

$$\bar{X}_{new} = \bar{X}_{old} - \bar{X}_{old}$$
$$= 0$$

while the standard deviation is unchanged. Next, each score obtained from the first step is divided by the original standard deviation. As a result,

$$\bar{X}_{new} = \frac{0}{\sigma_{old}} = 0$$

$$\sigma_{new} = \frac{\sigma_{old}}{\sigma_{old}} = 1$$

That is, the mean remains zero, while the standard deviation becomes 1. Summarizing these steps in a single formula gives

$$Z = \frac{X - \bar{X}}{\sigma}$$

where *Z* is the symbol for a standard score.

Converting each of the original examination scores given at the beginning of this chapter to Z scores yields:

	English	mathematics	psychology
X	80	65	75
$\bar{X}$	85	55	60
σ	10	5	15
Z	$\dfrac{80-85}{10}=-0.50$	$\dfrac{65-55}{5}=+2.00$	$\dfrac{75-60}{15}=+1.00$

Thus, the standard scores show at a glance that the student was half a standard deviation below the mean in English, two standard deviations above the mean in mathematics, and one standard deviation above the mean in psychology.

Table 6.1 *Hypothetical distribution of 20 height scores expressed as raw scores (in.), Z scores, and raw scores (cm) measured from the top of a 36-in. desk*

person	height: raw score (in.)	height: Z score	height: raw score (cm) measured from top of 36-in. desk
1	72	+2.36	91.44
2	70	+1.24	86.36
3	70	+1.24	86.36
4	70	+1.24	86.36
5	69	+0.67	83.82
6	69	+0.67	83.82
7	68	+0.11	81.28
8	68	+0.11	81.28
9	68	+0.11	81.28
10	68	+0.11	81.28
11	67	−0.45	78.74
12	67	−0.45	78.74
13	67	−0.45	78.74
14	67	−0.45	78.74
15	67	−0.45	78.74
16	67	−0.45	78.74
17	66	−1.01	76.20
18	66	−1.01	76.20
19	66	−1.01	76.20
20	64	−2.14	71.12
$\bar{X}$	67.80	0.00	80.77
σ	1.78	1.00	4.52

When raw scores are transformed into Z scores, *the shape of the distribution remains the same.* We just measure from a new point (the mean instead of zero), with a new unit size (the standard deviation instead of the raw units). To illustrate this point, a set of hypothetical heights in inches for 20 men is presented in Table 6.1 along with the corresponding Z scores. The scores have been arranged in increasing order for clarity. The mean of the raw height scores is 67.80 in. and the standard deviation is 1.78 in., and the Z scores were obtained using the formula given above; for example, the Z score corresponding to 72 in. is equal to

$$Z = \frac{72 - 67.80}{1.78} = +2.36$$

In accordance with the previous discussion, the mean of the Z scores turns out to be 0.00 and the standard deviation is 1.00. The raw score and Z score distributions are plotted in Figure 6.1; note that a single graph suffices because the shape of the distribution is the same for both sets of scores. Thus, the relationship of the scores to one another is *not* changed by transforming to standard scores; all that change are the location and the scaling.

Although the height scores are obtained by measuring the distance from the floor to the top of each individual, identical Z scores

Figure 6.1 *Frequency distribution of raw scores and Z scores in Table 6.1.*

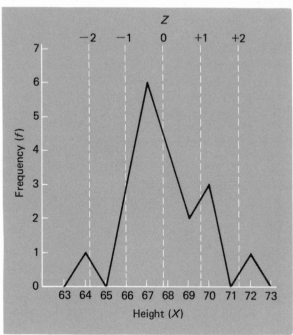

would be obtained if every person were measured from the top of a desk that is 36 in. high instead of from the ground. Even a change of scale to centimeters would not affect the Z scores, as is shown in the last column of Table 6.1. For example,

$$\frac{70-67.80}{1.78} = +1.24 = \frac{86.36-80.77}{4.52}$$

| in. | Z score | cm measured from top of 36-in. desk |

This indicates that Z scores give an accurate picture of the standing of each score relative to the reference group no matter where the original scores are measured from or what scale is used.

Standard scores are used extensively in the behavioral sciences. They are also related to certain statistics which play an important role in statistical inference, as we will see in subsequent chapters.

T SCORES

Standard scores have one disadvantage: they are difficult to explain to one not well versed in statistics. A college professor, in a stroke of genius that turned out to be highly misguided, once decided to report the results of an examination as Z scores. He was quickly besieged by anxious students who did not understand that a Z score of 0 represents average performance (and not zero correct!), not to mention the agitation of those who received negative scores and wondered how they could ever repay the points they owed the professor.

Since one function of behavioral scientists is to report test scores to people who are not statistically sophisticated, several alternatives to Z scores have been developed. The mean and standard deviation of each such "common denominator" have been chosen so that all scores will be positive, and so that the mean and standard deviation will be reasonably easy to remember. One such alternative, called T scores, is defined as a set of scores with a mean of 50 and a standard deviation of 10. The T scores are obtained from the following formula:

$$T = 10Z + 50$$

That is, each raw score is converted to a Z score, each Z score is multiplied by 10, and 50 is added to each resulting score. For example, a height score of 69 in Table 6.1 is converted to a Z score of $+0.67$ by the usual formula. Then, T is equal to $(10)(+0.67)$ $+50$ or 56.70. It is easy to show that the above formula does in fact produce the desired mean and standard deviation:

procedure	new $\bar{X}$	new σ
1. Convert raw scores to Z scores	0	1
2. Multiply each Z score by 10	$10 \times 0 = 0$	$10 \times 1 = 10$
3. Add 50 to each score obtained in Step 2	$0 + 50 = 50$	10 (no change)

Since the mean of T scores is 50, you can still tell at a glance whether a score is above average (it will be greater than 50) or below average (it will be less than 50). Also, you can tell how many standard deviations above or below average a score is; for example, a score of 40 is exactly one standard deviation below average (equivalent to a Z score of -1.00) since the standard deviation of T scores is 10. A negative T score is mathematically possible but virtually never occurs in practice; it would require that a person be over five standard deviations below average, and scores more than three standard deviations above or below the mean almost never occur with real data.

SAT SCORES

Scores on some nationally administered examinations, such as the Scholastic Aptitude Test (SAT), the College Entrance Examination Boards, and the Graduate Record Examination, are transformed to a scale with a mean of 500 and a standard deviation of 100. These scores, which we will call SAT scores for want of a better term, are obtained as follows:

$$SAT = 100Z + 500$$

That is, the raw scores are first converted to Z scores, each Z score is multiplied by 100, and 500 is added to each resulting score. The proof that this formula does in fact yield a mean of 500 and a standard deviation of 100 is similar to that involving T scores. (In fact, an SAT score is just ten times a T score.) This explains the apparent mystery of how you can obtain a score of 642 on a test with only several hundred items. And you may well be pleased if you obtain a score of 642, since it is 142 points or 1.42 standard deviations above the mean (and therefore corresponds to a Z score of $+1.42$ and a T score of 64.2).

SUMMARY

In order to compare scores based on different means and standard deviations, it is desirable to convert the raw scores to *transformed scores* with a common mean and standard deviation. Some frequently used transformed scores are Z scores (*standard scores*), with a mean of 0 and a standard deviation of 1 ; T scores, with a mean of 50 and a standard deviation of 10 ; and "*SAT scores*," with a mean of 500 and a standard deviation of 100. A transformed score shows at a glance whether a score is above or below average, and how far in terms of standard deviation units, with respect to a specified reference group.

APPENDIX TO CHAPTER 6: PROOFS OF RULES FOR CHANGING $\bar{X}$ AND σ

Rule 1. If a constant (positive or negative), k, is added to every score,

$$\bar{X}_{new} = \bar{X}_{old} + k$$

PROOF: Each new score may be represented by $X + k$. The mean of these scores, obtained in the usual way by summing and dividing by N, is

$$\bar{X}_{new} = \frac{\sum (X + k)}{N}$$

$$= \frac{\sum X + Nk}{N} \quad \text{(Rule 6, Chapter 1)}$$

$$= \frac{\sum X}{N} + \frac{Nk}{N}$$

$$= \bar{X}_{old} + k$$

Rule 2. If a constant (positive or negative), k, is added to every score,

$$\sigma^2_{new} = \sigma^2_{old}$$
$$\sigma_{new} = \sigma_{old}$$

PROOF: Each new score may be represented by $X + k$, and the mean of the new scores is $\bar{X}_{old} + k$. Therefore, the definition formula for the variance of the transformed scores becomes

$$\sigma^2_{new} = \frac{\sum ([X+k] - [\bar{X}_{old}+k])^2}{N}$$

$$= \frac{\sum (X+k-\bar{X}_{old}-k)^2}{N}$$

$$= \frac{\sum (X-\bar{X}_{old})^2}{N}$$

$$= \sigma^2_{old}$$

And since $\sigma^2_{new} = \sigma^2_{old}$, $\sigma_{new} = \sigma_{old}$.

Rule 3.   If every score is multiplied by a constant k (positive or negative, and greater than or less than one),

$$\bar{X}_{new} = k\bar{X}_{old}$$

PROOF: Each new score may be represented by kX, and the mean of these scores is

$$\bar{X}_{new} = \frac{\sum kX}{N}$$

$$= k\frac{\sum X}{N} \quad \text{(Rule 8, Chapter 1)}$$

$$= k\bar{X}_{old}$$

Rule 4.   If every score is multiplied by a positive constant k,

$$\sigma^2_{new} = k^2\sigma^2_{old}$$

$$\sigma_{new} = k\sigma_{old}$$

PROOF: Each new score may be represented by kX, and the mean of the new scores is $k\bar{X}_{old}$. The variance is equal to

$$\sigma^2_{new} = \frac{\sum (kX - k\bar{X}_{old})^2}{N}$$

$$= \frac{\sum (k[X - \bar{X}_{old}])^2}{N}$$

$$= \frac{\sum k^2(X - \bar{X}_{old})^2}{N}$$

$$= \frac{k^2 \sum (X - \bar{X}_{old})^2}{N}$$

$$= k^2\sigma^2_{old}$$

$$\sigma_{new} = \sqrt{k^2\sigma^2_{old}}$$

$$= k\sigma_{old}$$

part III inferential statistics

7 the general strategy of inferential statistics

A distressing fact that complicates the life of the behavioral scientist is that the populations from which he seeks to collect his data are usually far too large to permit the measurement of every element in the population—or even of a sizable portion of the population. For example, a psychologist who wishes to test the hypothesis that the use of programmed learning techniques will improve the performance of students in statistics courses is confronted with the population of all college students in the United States who are studying statistics. This is enough to frighten off even the most dedicated researcher, for he cannot invest the time, effort, and financial resources needed to obtain and analyze data from so many thousands of people. In desperation, the psychologist might decide to limit the study to 100 statistics students selected in some way from the population. This procedure will provide him with a group that can be measured in its entirety, but a new and quite serious problem is created. Since the data are obtained from only a very small part of the population, it is certainly possible that the results of the experiment would not apply at all to a different group of 100 students, or—more importantly—to the entire population. If this is in fact the case, the unhappy consequence will be that the results published by the psychologist, and his decision concerning the merits of programmed learning procedures, will subsequently be contradicted by other researchers who attempt to verify his findings by using different groups of students drawn from the population. This will result in professional loss of face for the psychologist, and, more importantly, a setback in the attempt to further the development of scientific knowledge in this area while the various conflicting research results are untangled and attempts made to sort out the true state of affairs. Unfortunately, the only way to be certain of preventing this calamity would be to measure the entire population. Thus, the behavioral scientist needs to draw his conclusions about the population, but can usually measure only a small part of that population—which would seem, on the surface, to be an insoluble problem.

This unpleasant problem is typical in all areas of the behavioral sciences. For example, an industrial psychologist studying the effects of pay on job satisfaction cannot possibly measure the entire population of all paid employees in the United States. Even if he were to restrict his attention to one type of job, the resulting population would still be far too large to measure in its entirety. An experimental psychologist studying the performance of rats in a maze under varying conditions cannot run all laboratory rats in the world in his experiment, nor can he obtain all possible runs from any one rat. An anthropologist interested in the effects of different cultures on motivation in children cannot study all of the children from each of the cultures in which he is interested. In fact, this predicament is so common that most behavioral science research would not be possible without some effective resolution of the problem. As you may have guessed from the title of this chapter, the answer lies in the use of *inferential statistics*—techniques for drawing inferences about an entire population based on data obtained from a sample drawn from the population.

THE GOALS OF INFERENTIAL STATISTICS

The kinds of inferences that the behavioral scientist wishes to make about the population can be usefully subsumed under three related categories. One procedure of interest is to estimate the values of population parameters. Given a sample whose mean and variance have been computed, he may wish to estimate the mean and variance of the parent population. Any statistic computed from a single sample, such as $\bar{X}$ or s, which provides an estimate of the corresponding population parameter, such as μ or σ, is called a *point estimate*.

For some purposes, a point estimate is not sufficiently informative, since it leaves us with no idea of how far off from the parameter it may be. Such information is supplied by an *interval estimate*, a range of values which has a known probability of including the true value of the parameter. For example, suppose that a sample of white mice runs through a maze in an average of 60 seconds. A point estimate would state that the most reasonable value of μ is 60. If an estimate in terms of a single number is *not* essential, the researcher may instead decide to use appropriate techniques of inferential statistics to determine an *interval* which is likely to include the population mean. Thus, he might find that it would be reasonable to expect the population mean to fall in the interval 54–66, or 60 ± 6. The interval estimate divides all numerically possible values into two sets: "likely" (54–66), and "unlikely" (less than 54 and more than 66), with "unlikely" having a precise meaning to be discussed in subsequent sections.

A third important use of inferential statistics is to assess the probability of obtaining certain kinds of sample results under certain population conditions. For example, an educational researcher might wish to determine the probability that a sample of 100 statistics students, who have been switched to programmed learning techniques and have gained an average of 5 points in their statistics examination scores, come from a population where the mean gain in examination scores is zero. In other words, if the programmed learning techniques do *not* really work on the average for the *population*, how likely is it that we can get a *sample* of 100 statistics students whose mean change is 5 points?

THE STRATEGY OF INFERENTIAL STATISTICS

To illustrate the general strategy of inferential statistics, let us look at an example, based on games of chance, that has a great many similarities to the research paradigm. To keep the computations as simple as possible, we will choose a straightforward game—that of flipping coins with a friend. Suppose that you wager $1 on the result of the toss of a coin, your friend calls "heads," and the coin lands heads up. You are now out one dollar, but resolve to recoup your loss on the next toss. Unfortunately, your friend continues to call "heads," the coin lands heads up six times in a row, and you lose six consecutive times. At this point, you begin to entertain some suspicions about the possibility of a biased coin (especially if your friend insists on being the one to call the toss). If the coin is honest, you would like to continue to play and try and regain some of the $6 you have lost; but if your friend has foisted off a biased coin on you, you should terminate the game and denounce your friend as dishonest.* We can summarize these possibilities in the form of experimental hypotheses:

HYPOTHESIS 0:   The results are due to chance. The coin is honest, and your friend has simply had a run of luck.

* For simplicity, we will assume that other possible alternatives, such as mutual agreement to use a new coin, have been rejected by the contestants.

HYPOTHESIS 1 :   The coin is biased (in favor of falling heads up).

Needless to say, it makes quite a bit of difference as to what decision you choose, as can be seen from the following table comparing your decision with what the true state of affairs might be :

| | | true state of affairs or state of the population | |
		Hypothesis 0 is true; results are due solely to chance	Hypothesis 1 is true; coin is biased
your decision	do not reject Hypothesis 0 (keep playing)	correct decision	error—continued financial support of dishonest friend
	reject Hypothesis 0 (quit game and denounce friend)	error—unfair slander of honest friend	correct decision

The only way of ensuring an exactly correct decision in this touchy situation would be to measure the entire population. In order to do this, it would be necessary to flip the coin an infinite number of times. If the coin were ultimately found to fall heads up more than half the time, your friend would indeed have an unfair advantage ; while if the coin came up heads half the time and tails half the time, the game would be fair. Unfortunately, even if generations of people did nothing but flip the accursed coin during every waking moment of their lives, the task of registering an infinite number of tosses would not be finished until the end of time—so you can hardly afford to wait around until the results of this "experiment" are in.

A better procedure is to draw a sample of tosses from the population. In practice, a fairly large sample should be obtained (perhaps as many as 100 flips of the coin if your will power survives that long), but for purposes of this illustrative example, let us assume that you must reach a decision based only on the six results you have already observed. How can you use the information from this sample to draw inferences about whether or not the coin is in fact biased ? In order to be able to answer this question, you must first learn some basic definitions and computations regarding the probability of a given event.

PROBABILITY

Definition of probability. The *probability* (P) of a given event is defined by the following fraction:

$$P(\text{event}) = \frac{\text{number of ways the specified event can occur}}{\text{total number of possible events}}$$

For example, if an unbiased coin is flipped and you wish to determine the probability that heads will come up, the solution is

$$P(H) = \frac{1}{2} = \frac{1 \text{ head}}{2 \text{ possible events: head or tail}}$$

Similarly, the probability of "tails up" is also $\frac{1}{2}$. Decimals are also often used to express probabilities, so we can also write $P(H)$ (that is, the probability of obtaining a head on one toss of the coin), or $P(T)$, as .50. These numbers simply express the fact that if the coin is fair, heads will occur half the time and tails will occur half the time.

To provide some further (and less obvious) illustrations, let us assume that you have a standard 52-card deck of playing cards.* The deck is thoroughly shuffled, and one card is drawn. The probability that it is the king of diamonds is 1/52 (or .019); there is only one way for that event to occur (only one king of diamonds) and 52 possible equally likely events (cards that could be drawn). In fact, the probability of drawing any one specific card from the deck is 1/52. The probability of drawing a nine is 4/52 or 1/13 (or .077), since there are four ways for this event to occur (nine of spades, nine of hearts, nine of diamonds, and nine of clubs) out of the total possible 52 events. If we define a "spot card" as anything below a ten, the probability of drawing a spot card in clubs is 8/52, or 2/13 (or .15), for there are eight specific equally likely events (the two, three, four, five, six, seven, eight, and nine of clubs) in the 52-card deck.

Probability cannot be less than 0 or greater than 1. If an event has probability zero, that means that it cannot possibly happen; if an event has probability 1, that means that it must happen. Thus, for example, the probability of drawing either a red or black card from a standard deck is 1.00, since all the cards in the deck are either red or black (that is, $P = 52/52 = 1.00$). The probability of drawing a purple card with yellow polka dots is zero ($P = 0/52 = 0$).

---

* The standard deck contains 52 cards divided into four suits (spades, hearts, diamonds, and clubs), each of which contains thirteen cards (two through ten, jack, queen, king, and ace).

Odds. The odds against an event are defined as the ratio of the number of unfavorable outcomes to the number of favorable outcomes, all outcomes being equally likely.

EXAMPLE:

outcome	total events (T)	number of ways of obtaining the event (W)	number of ways of not obtaining the event $(T-W)$	probability of event (W/T)	odds against event $(T-W$ to $W)$
king of diamonds	52	1	51	1/52	51 to 1
a nine	52	4	48	4/52 or 1/13	48 to 4 or 12 to 1
spot card in clubs	52	8	44	8/52 or 2/13	44 to 8 or 11 to 2 ($5\frac{1}{2}$ to 1)

If the odds against your winning a game are (say) 12 to 1, you need to collect 12 times your wager when you do win for the game to be fair. If you stand to win more than 12 times your bet, it is to your advantage to play the game; and if you collect less than 12 times your bet when you do win, you should refuse to play because you figure to lose in the long run.

The probability of A or B. The probability of *either of two events* taking place is conveniently computed by means of the following formula:

$$P(A \text{ or } B) = P(A) + P(B) - P(A \text{ and } B)$$

EXAMPLE: Suppose that you will win a prize if you draw *either* a king *or* a club from a standard 52-card deck in a single try. The probability of winning is equal to

$P(\text{king}) + P(\text{club}) - P(\text{king and club})$
$= 4/52 \quad + 13/52 \quad - 1/52$
$= 16/52 \text{ (or .31)}.$

To verify the above calculations, let us tackle the problem the long way. There are the usual 52 total possible events, and the number of favorable events may be counted as follows:

king (*4 favorable* *events*)	*club* (*13 favorable* *events*)
king of spades	ace of clubs
king of hearts	*king of clubs*
king of diamonds	queen of clubs
king of clubs	jack of clubs
	ten of clubs
	nine of clubs
	eight of clubs
	seven of clubs
	six of clubs
	five of clubs
	four of clubs
	three of clubs
	two of clubs

There appear to be 17 favorable events—13 clubs and 4 kings—but the king of clubs has been counted twice, and there is only one king of clubs in the deck. Even a king does not deserve such special treatment, especially since it will render the final calculations incorrect. The $P(A \text{ and } B)$ term is subtracted to allow for events which have been counted twice. Therefore, the probability of king *and* club (1/52, since there is one king of clubs in the deck) is subtracted from the previous figures, and the resulting probability (16/52) shows that there are 16 separate and distinct favorable events (cards) out of the total of 52 in the deck.

If events A and B cannot happen simultaneously (in mathematical terminology, are *mutually exclusive*), then $P(A \text{ and } B) = 0$. For example, the probability of drawing a king or a queen from the deck is equal to $4/52 + 4/52 - 0/52 = 8/52$. The events are mutually exclusive inasmuch as no card in the deck is *both* a king and a queen.

The probability of A and then B.   Suppose that a card is drawn from a standard deck, looked at, and replaced in the deck. The deck is shuffled thoroughly, and a second card is drawn. We wish to know the probability of obtaining a king on the first draw *and then* the ace of spades on the second draw. We assume that the two draws are *independent*—that is, what happens on the first draw does not affect the probabilities on the second draw (because of the shuffling). The solution is given by the formula

$$P(A \text{ and then } B) = P(A) \times P(B)$$

In the present example, the answer is

$$P(\text{king and then ace of spades}) = P(\text{king}) \times P(\text{ace of spades})$$
$$= 4/52 \times 1/52$$
$$= 4/2704 \text{ (or .0015)}$$

If the card drawn on the first try is *not* returned to the deck, the probability becomes $4/52 \times 1/51$ (or 4/2652). Under this procedure, only 51 cards remain in the deck after the first draw, and the probability of success on the second draw is therefore 1/51.

To return to the coin problem, the probability of obtaining a head in each of two consecutive flips of a coin is $P(H) \times P(H)$, which is equal to $1/2 \times 1/2$ or 1/4. This can easily be verified by listing all possible outcomes of two flips of the coin:

first flip	second flip	overall result	probability
H	H	2 heads	1/4 (or .25)
H	T	1 head, 1 tail	1/4
T	H	1 head, 1 tail	1/4
T	T	2 tails	1/4

Of the four possible outcomes all of equal probability, only one yields two heads; hence the probability of two heads is 1/4. Similarly, the probability of two tails is 1/4, and the probability of obtaining one head and one tail irrespective of order is 2/4 or 1/2. Since the events are independent (that is, the results of the first flip do not affect the probabilities on the second flip), the probabilities are exactly the same if we flip two coins once.

CONCLUSION OF THE COIN "EXPERIMENT"

Much more could be said on the topic of probability, but we will leave further discussion of this interesting area for the mathematicians and return to the problem concerning flipping coins with a friend. In your quest to determine whether or not the coin is biased, it might well be helpful to know the probability of obtaining six consecutive heads. After all, you would have little reason to accuse your friend of unethical practices if this were a common, everyday occurence, and you would be best advised simply to continue the game and hope to benefit from a run of tails. On the other hand, if six heads in a row were a "million-to-one" shot ($P = 1/1,000,000$ or .000001), you might well be skeptical about the reason for your losses. The best strategy in such a case might well be to assume that bias in the coin rather than a spectacularly lucky event was responsible for the results.

If you have been thinking ahead, you may have observed that the rule for determining P(A and then B) would appear to yield information that would be helpful in the present dilemma. That is, you might wish to make use of the fact that

$$P(\text{six consecutive heads}) = P(\text{head on flip \#1}) \times P(\text{head on flip \#2})$$
$$\times\, P(\text{head on flip \#3}) \times P(\text{head on flip \#4})$$
$$\times\, P(\text{head on flip \#5}) \times P(\text{head on flip \#6})$$

Unfortunately, this formula requires that you know the probability of a head on any one flip, and this probability is equal to 1/2 *only if the coin is unbiased.* If the coin is "loaded," the true probability of a head might be anything other than 1/2; $P(H)$ could equal 2/3, 3/4, or even 1 (if the coin were two-headed). Thus, you would like to know the probability of obtaining six consecutive heads in order to determine whether the coin is biased—that is, more likely to fall heads up on any given trial. However, you need to know the probability that the coin will fall heads up on a given trial to determine the probability of six consecutive heads.

To extricate yourself from this predicament, you must use the only certain fact at your disposal—that *if the coin is fair, the probability of a head on one trial is 1/2.* This is done as follows:

Step 1.   Assume that the coin is fair (that $P(H) = 1/2$).

Step 2.   Determine the probability of six heads, basing the calculations on the assumption stated in Step 1.

Step 3.   The calculations in Step 2 will yield the probability of six consecutive heads *if* the coin is fair. You may then choose one of two decisions:

Either

A.   Based on the probability computed in Step 2, you decide *not* to reject the assumption that the coin is fair;

or

B. The probability computed in Step 2 is so small that the assumption in Step 1 is best regarded as incorrect. While such an event could occur, it is so unlikely that it is better to conclude that the assumption that $P(H) = 1/2$ is wrong, so you reject this assumption and conclude that the coin is *not* fair.

You can now complete the experiment. *Assuming that the coin is fair*, the probability of six heads is equal to $(1/2)^6$ or

$$1/2 \times 1/2 \times 1/2 \times 1/2 \times 1/2 \times 1/2 = 1/64 \qquad \text{or} \qquad .016$$

Your final task is to choose between two decisions. Either the coin is fair, and an event with probability .016 has occurred (Hypothesis 0), *or* the coin is unfair, $P(H)$ *is not equal to 1/2, and therefore P (six consecutive heads) is not equal to .016* (Hypothesis 1). Remember that the value of .016 is the correct probability for six consecutive heads only if the coin is fair and $P(H) = 1/2$. If this assumption is abandoned, the probability of six heads in a row is by no means necessarily .016. (For example, if the coin is so drastically biased that it always falls heads up, as would happen if it were a two-headed coin, $P(H) = 1.00$ and P (six consecutive heads) $= (1.00)^6 = 1.00$. Less drastically, if for that coin $P(H) = .75$, then P (six consecutive heads) $= (.75)^6 = .18$.) Behavioral scientists commonly follow the arbitrary (but reasonable) practice of regarding events with probabilities less than .05 as "sufficiently unlikely to justify rejecting Hypothesis 0," and events with probabilities greater than .05 as "not sufficiently unlikely" to justify such rejection.* Using this rule, you would abandon the assumption that the coin is fair and terminate the game; rather than assume that an "unlikely" event had occurred (six consecutive heads with a fair coin), you would instead conclude that the coin was biased in favor of falling heads up.

Now that you have reached a decision, the "experiment" is completed. The statistical analysis cannot guarantee a correct

---

* Since this is a matter of convention and is not based on any mathematical principle, common sense is essential in borderline cases. If, for example, $P = .06$, the technically correct procedure is to call the event "not sufficiently unlikely," but .06 is so close to .05 that the best plan is to repeat the experiment and gain additional evidence. The reason that a guide such as the ".05 rule" is needed is that *you must reach a decision*—here, either to keep playing or to stop the game. Also, some researchers use a more stringent criterion for a judgment of "unlikely," namely .01; and others use these two and other criteria (such as .10) flexibly, depending on the circumstances of the research.

decision (remember, you cannot ensure that you are right because you are unable to measure the whole population), and it is still possible that you have made an error (here, that you are being unfair to a friend who is actually honest). However, statistics will point the way to the "best bet"—the decision about the population that is most warranted by the results obtained from the sample.

THE COIN EXPERIMENT AND BEHAVIORAL SCIENCE RESEARCH COMPARED

The analysis of a behavioral experiment by means of inferential statistics proceeds in an almost identical fashion to the analysis involving the flipping of the coins. Thus, if you understand the rationale underlying the coin experiment, you should have little trouble comprehending the use of inferential statistics in the behavioral research experiment. For example, let us return to the educational researcher who is hopeful of demonstrating gains in students' knowledge of statistics if techniques of programmed learning are applied. Since it is impossible to measure the entire population, he obtains a sample of 100 statistics students chosen in such a way as to avoid bias insofar as possible. The students are given the programmed learning materials, and a 5-point increase in the average statistics examination score of the group results. The psychologist applies the appropriate inferential statistic and finds, to his regret, that the probability of such an increase assuming that the programmed learning techniques do *not* work is .24. He therefore does *not* reject Hypothesis 0—the hypothesis that programmed learning does *not* improve knowledge of statistics insofar as the entire population of statistics students is concerned—because a 5-point increase in average examination scores in the sample *is* reasonably likely if the mean increase in the population is zero.

·The parallel between the coin experiment and the behavioral research paradigm is summarized in Table 7.1.

The techniques discussed in the remainder of this book are procedures for computing (a) point estimates, (b) interval estimates, and (c) probability values such as the .24 in the hypothetical programmed learning experiment. The reason that the rest of this book has so many pages is that different kinds of inferential statistics are needed for different kinds of experiments designed to answer different kinds of questions.

Table 7.1 *Coin experiment and behavioral research compared*

	coin experiment	*behavioral research* (*programmed learning experiment*)
Population	Infinite number of flips of coin	Vast number of people (Statistics students)
Sample	Smaller, more feasible number of flips	Smaller, more feasible number of people
Hypothesis 0	Assume that coin is fair, and that any deviation from exact 50–50 heads–tails split in sample is due to chance	Assume that programmed learning is *not* useful, and that any deviation from exact zero improvement in sample is due to the particular cases that happened to fall in the sample, that is, chance
Hypothesis 1	Results are *not* due to chance (coin is biased)	Results are *not* due to chance (programmed learning *is* useful)
Experiment	Obtain data (flip coin 6 times—result: 6 heads)	Obtain data (measure sample—result: 5 point mean increase)
Statistical analysis	Compute P (six consecutive heads)	Compute P (five point increase)
Results of statistical analysis	P (six consecutive heads) *if* coin is fair (answer: .016)	P (5 point increase) *if* programmed learning is *not* beneficial (answer: .24)
Decision if P is "not sufficiently small" using ".05 rule" ($P = .051–1.00$)	Do *not* reject Hypothesis 0. The assumption that the coin is fair is a reasonable one. Continue playing. *Risk*: Continued financial support of dishonest friend. (Size of risk not conveniently determined. See Chapter 13.)	Do *not* reject Hypothesis 0. The assumption that programmed learning is *not* useful is a reasonable one. Retain standard teaching methods. *Risk*: Failure to adopt new method that is actually better. (Size of risk not conveniently determined. See Chapter 13.)

Table 7.1 (*Continued*)

	coin experiment	behavioral research (programmed learning experiment)
Decision if P is "sufficiently small" using ".05 rule" (P = .00–.05)	Reject Hypothesis 0; accept Hypothesis 1. The assumption that the coin is fair is *not* reasonable. Terminate game. Risk: Unfair slander of honest friend (size of risk ≤P).	Reject Hypothesis 0; accept Hypothesis 1. The assumption that programmed learning is not beneficial is *not* reasonable. Switch to pro- grammed learning. Risk: Adoption of new method that is no better or worse than standard teaching methods (size of risk ≤P).
Actual decision	Reject Hypothesis 0. Terminate game. Since hypothesis 0 has been rejected, we no longer need assume that an event with probability .016 has occurred.	Do not reject Hy- pothesis 0. It cannot be concluded that programmed learning is beneficial under these circumstances.

STATISTICAL MODELS

Suppose that in the coin experiment, *five* heads had been obtained in six flips (and your friend is therefore ahead $4). What decision should be reached about the possibility of a biased coin in this case? To compute this probability, it is necessary to enumerate all possible ways of obtaining five heads in six tosses* :

flip : 1	2	3	4	5	6
A H	H	H	H	H	T
B H	H	H	H	T	H
C H	H	H	T	H	H
D H	H	T	H	H	H
E H	T	H	H	H	H
F T	H	H	H	H	H

* Mathematics maкe use of a special technique called "permutations" to enumerate the various possibilities. Since a knowledge of permutations is not necessary for comprehension of the material in this book, it will be omitted.

As you may have guessed from the fact that the probability of six consecutive heads is 1/64, there are a total of 64 possible and equally likely outcomes if a fair coin is flipped six times. Thus, the probability of any one of the outcomes shown above is 1/64, and the total probability of five heads in six flips is equal to

P(A or B or C or D or E or F).

Since the events are obviously mutually exclusive, this equals $1/64 + 1/64 + 1/64 + 1/64 + 1/64 + 1/64 = 6/64$ or .094.

Using the same procedures, we can similarly enumerate the ways and then compute the probability of any number of heads if a coin is flipped six times. The results are as follows:

number of heads	probability
0	.016
1	.094
2	.234
3	.312
4	.234
5	.094
6	.016

This *statistical model* gives the probabilities of events (number of heads) under stated conditions (that $P(H) = 1/2$). If an observed event or sample result is highly improbable or unlikely under the model, then the assumptions about the conditions may need revision. Thus, when you obtained six heads in the previous experiment, this result was very unlikely in terms of the statistical model ($P = .016$), and you therefore conclude that the assumption underlying the model (that $P(H) = 1/2$) was incorrect—that is, that the coin was biased. If instead you obtained five heads in six flips, you would not reject the assumption that the coin was fair. The probability of obtaining five or more heads if $P(H) = 1/2$ is equal to $.094 + .016$ or .11, and this is a "not sufficiently unlikely" result in terms of the ".05 rule." It is particularly "not sufficiently unlikely" if the question is, "What is the probability of 5 or more heads *or* 5 or more tails" for which the answer is $.11 + .11 = .22$. Similarly, the model presents information about any other possible outcome; and we could also define the models for games involving different numbers of flips in a similar fashion.

Unfortunately, it is more difficult to determine statistical models in problems in the behavioral sciences. In the coin experiment, it is evident that there is one favorable event (head) and a total of two possible and equally likely events (head or tail), and the probability of obtaining a head on one flip of a fair coin is easily defined as 1/2. But what of the programmed learning experiment and 5-point increase in examination scores? Defining "favorable events" and "total possible events" in this situation—or in any research experiment—is by no means a simple task, yet a statistical model is essential in order to provide the probability values that the experimenter needs to reach a decision.

One possible (but not practical) way in which the necessary statistical model could be obtained would be to draw a large number of samples of the same size from the defined population and determine on an empirical basis how often the various alternatives occur. The samples should be *random**—that is, drawn in such a way as to minimize bias and make the sample typical of the population insofar as possible. If the mean of each sample is computed and a frequency distribution of the many means is plotted, we will have an idea how often to expect a sample with a particular mean. Such a frequency distribution is called an *experimental sampling distribution* because it is obtained from observed or experimental data.† Note that this is not how distributions of sample values are ordinarily determined, but merely how one *might* determine them.

As an example, suppose that we draw a random sample of 100 American men, measure the height of each man in the sample, and compute the mean height of the sample. We then draw another random sample of 100 American men and compute the mean height of this sample. In a startling display of perseverance, we obtain a total of 1000 samples (each of $N = 100$). Table 7.2 shows the frequency distribution of mean heights of the 1000 samples, and Figure 7.1 graphically represents the frequency distribution.

The experimental sampling distribution serves as the statistical model for assigning probabilities to any "height" observation, and Column C of Table 7.2 serves the same function as the table of probabilities of any number of heads in six flips of a coin discussed previously. The probability of obtaining a sample of a particular mean height is thus defined as the number of *empirically determined* favorable events to the number of *empirically determined* total events. That is, if in 1,000 samples a mean height in the interval

---

* A *random sample* is defined as a sample drawn in such a way as to (1) give each element in the population an equal chance of being drawn *and* (2) make all possible samples of that size equally likely to occur.
† It is also sometimes called a Monte Carlo distribution, particularly when generated by means of an electronic computer.

Table 7.2 *Hypothetical frequency distribution of mean height for 1000 samples (sample N = 100)*

(A) mean height (*in.*)	(B) number of samples (*f*)	(C) proportion of samples ($p = f/N$)
68.55 or more	0	.000
68.50–68.54	1	.001
68.45–68.49	1	.001
68.40–68.44	4	.004
68.35–68.39	11	.011
68.30–68.34	19	.019
68.25–68.29	31	.031
68.20–68.24	56	.056
68.15–68.19	81	.081
68.10–68.14	112	.112
68.05–68.09	116	.116
68.00–68.04	141	.141
67.95–67.99	131	.131
67.90–67.94	103	.103
67.85–67.89	76	.076
67.80–67.84	57	.057
67.75–67.79	28	.028
67.70–67.74	21	.021
67.65–67.69	7	.007
67.60–67.64	2	.002
67.55–67.59	2	.002
67.50 or less	0	.000
Total	1000	1.000

67.90–67.94 occurs 103 times, then the probability that *any one* sample of size $N = 100$ that we might draw will have a mean height between 67.90 and 67.94 is 103/1000 or .103. (Also, if we were to obtain many other random samples of $N = 100$ from this population, we would expect approximately 10% of the samples to have a mean between 67.90 and 67.94.) The probability of drawing a sample with a mean height of 68.25 *or more* would equal .031 + .019 + .011 + .004 + .001 + .001, or .067. Other probabilities could be determined in a similar fashion.

The above procedure for determining an experimental sampling distribution is so laborious that it hardly represents any gain at all over measuring the entire population; it was presented so that you could visualize and understand the result. In practice, you will use

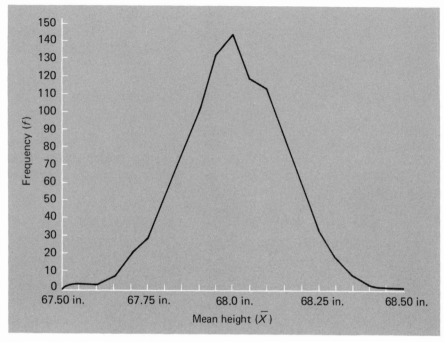

Figure 7.1 *Hypothetical distribution of mean height for 1000 samples* (*sample N = 100*).

theoretical sampling distributions. In many instances you will be able to benefit from work done by previous researchers and statisticians, and the form of the sampling distribution of the variable in which you are interested will be known in advance. For example, observe that in the experimental sampling distribution of height, most of the sample means cluster about the grand mean of 68.00 and there are fewer and fewer observations as one goes further away from the grand mean. This distribution closely approximates that of the theoretical *normal curve*, a statistical model which describes the nature of the sampling distributions of several statistics of interest, including means. Use of the theoretical normal curve model in the proper situations will enable you to avoid the toil and trouble of determining an experimental sampling distribution in order to ascertain the needed probabilities.

For example, the sampling distribution of means is known to have the shape of a normal curve for large samples. Therefore, you can simply proceed on the assumption that if you actually were to take a great many random samples of a given size from the population, the means of the samples would be normally distributed about the population mean (μ), just as they were in the height illustration. By making this assumption, probabilities are assigned on the basis of the theoretical sampling distribution, and an experimental or observed distribution of sample means is no longer necessary. You can then draw just one sample and compare the results to the theoretical model. In addition to means, this strategy can also be used with other statistics, as we will see later. The major precaution that should be emphasized is that you must be careful to use the normal curve model only in situations where it is proper to do so, for the normal curve model is not the only model used to make inferences about the population. It *is* one of the most important, however, and the one with which the discussion of specific procedures in inferential statistics will begin.

SUMMARY

Behavioral scientists usually wish to *draw inferences about a population* that is too large to measure in its entirety *based on information obtained from a sample* drawn *randomly* from that population. An initial hypothesis, called "Hypothesis 0," is assumed to be true so that the statistical calculations can be performed; these calculations yield the probability of obtaining the observed result *if* Hypothesis 0 is true. Using this probability, Hypothesis 0 is *rejected* in favor of the alternative hypothesis (Hypothesis 1) if the results are "sufficiently unlikely" to occur if Hypothesis 0 is true; Hypothesis 0 is *retained* (*not rejected*) if the results are "not sufficiently unlikely" to occur if Hypothesis 0 is true. A summary of such probabilities for all possible outcomes in a given experimental situation is called a *statistical model*; since it is not feasible to develop such models empirically, *theoretical* models are used to decide whether to retain or reject Hypothesis 0.

8 the normal curve model

Having taken pains in the preceding chapter to justify the existence of statistical models, it is now time to examine one in detail. A statistical model that you will find extremely useful in behavioral research is the *normal distribution*. This model is a *theoretical* distribution, since its shape is defined by a mathematical equation* and would *never* be exactly duplicated by empirical data. There are several reasons why this model is so important:

1. For many variables, you can legitimately *assume* that the shape of the *population distribution* is normal. That is, *without measuring the population*, you can assume the score distribution to be as shown in Figure 8.1.

2. Most importantly, the *sampling* distribution of various statistics tends to be normal if the sample sizes are large. For example, if a great many random samples are drawn from a population, and all are of the same size and contain about 30 or more observations each, their means when organized into a frequency distribution will form a sampling distribution of approximately normal shape.†

Thus, Figure 8.1 may be used to represent either (1) individual *scores* in a normally distributed population (for example, the height of every adult male in the United States), or (2) a normal *sampling*

---

* The formula for the normal curve is fairly complicated:

$$f(X) = \frac{1}{\sigma\sqrt{2\pi}} \, e^{-(X-\mu)^2/2\sigma^2}$$

where $\pi = 3.1416$ and $e = 2.7183$, both approximately.

† In addition to these two uses of the normal curve, many problems in inferential statistics are difficult to solve if the "technically correct" theoretical distribution is used. Often, however, the normal curve is a good approximation of the true distribution and can be used in its place, making matters simpler with little practical loss in accuracy.

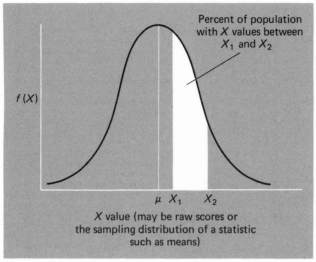

Figure 8.1   *The normal curve: a theoretical distribution.*

distribution of a statistic (for example, the *mean height* of each of 1000 large samples). This chapter deals with the normal distribution of individual scores in a population; drawing inferences about the mean of a population will be discussed in the next chapter.

SCORE DISTRIBUTIONS

As in the case of any frequency distribution, the horizontal axis of the normal distribution in Figure 8.1 represents score values and the vertical axis is related to frequency. Also, all values in Figure 8.1 are *under* the curve, so the *total population* is expressed in the graph as the *total area under the normal curve*; and we can arbitrarily define this area as 100% without bothering about the exact number of cases in the population.

Now suppose you are asked to find the proportion of the population with values between points X_1 and X_2 in Figure 8.1. Graphically, the answer is given by the percent *area between* these two points (which is equivalent to the percent frequency of these values and/or the probability of occurrence of these values).

If you have had an elementary calculus course, you may know that the numerical value of the area between these two X values may be found by integrating the normal curve equation between them. Panic may begin to set in at this point, for even students familiar with integral calculus would be properly horrified at the thought of carrying out an integration every time they wished to use the normal curve model. Fortunately, all necessary integrations have already been performed and summarized in a convenient table which shows the percent frequency of values within any distance (in terms of standard deviations) from the population mean. All that you are asked to do in order to use this table is to convert your raw scores to z scores.* (Otherwise, an infinite number of tables would be needed, one for each mean and standard deviation that a researcher might ever encounter.) This useful table is presented as Table B in the Appendix, and part of it is reproduced below.

	percents calculated from area under the normal curve					
z	.00	.01	.02	.03	...	.09
0.0						
⋮						
1.0	34.13	34.38	34.61	34.85		
1.1	36.43	36.65	36.86	37.08		
1.2	38.49	38.69	38.88	39.07		
1.3	40.32	40.49	40.66	40.82		
⋮						
∞						

The vertical column on the left represents z values expressed to one decimal place; the top horizontal column gives the second decimal place. The values within the table represent the *percent area between the mean and the z value* expressed to two decimal places. For example, to find the percent of cases between the mean and 1.03 standard deviations from the mean, select the 1.0 row and go across to the .03 column; the answer is read as 34.85%.

The table gives percents on one side of the mean only, so that the maximum percent given in the table is 50.00 (representing half of the area). However, the normal curve is *symmetrical*, which implies that the height of the curve at any positive z value (such as +1) is exactly the same as the height at the corresponding negative value (or −1). Therefore, the percent between the mean and a z of +1 is exactly the same as the percent between the mean and a z of −1, and presentation of a second half of the table to deal

---

* In the discussion of standard scores in Chapter 6, a capital Z was used to emphasize the fact that Z units do *not* change the shape of the original distribution and that the use of Z units does *not* require that the X distribution be normal. When the Z unit is used in conjunction with the normal curve, a lower case z will be used.

with negative scores is unnecessary. It is up to you to remember, therefore, whether you are working in the half of the curve above the mean (positive z) or the half of the curve below the mean (negative z). Also, do not expect the value obtained from the table to be the answer to your statistical problem in every instance. Since the table only provides values between the mean and a given z score, some additional calculations will be necessary whenever you are interested in the area between two z scores (neither of which falls at the mean) or the area between one end of the curve and a z score. These calculations will be illustrated below.

In Figure 8.2, the normal curve is presented with the X axis marked off to illustrate specified z distances from the mean. At the mean itself, the z value is 0 since the distance between the mean and itself is 0. The percent areas between the mean and other z values have been determined from Table B. Notice that the percent of the total area between μ (the population mean where $z = 0$) and a z of +1.00 is 34.13%, and the area between μ and $z = -1.00$ is also 34.13%. Thus, 68.26% of the total area under the curve lies between $z = -1.00$ and $z = +1.00$. Ninety-five percent of the total curve lies within approximately two standard deviation units of μ (more exactly, between $z = -1.96$ and $z = +1.96$). Thus, *if* the assumption of the normal curve model is correct, approximately 95% of all outcomes can be expected to fall between the mean and ±2 standard deviations. For example, if the heights of American adult males were assumed to be normally

Figure 8.2 *Normal curve: percent areas from the mean to specified z distances.*

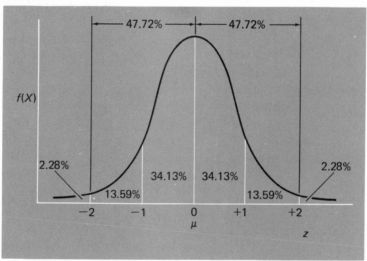

distributed with the mean at 5 ft 8 in. and the standard deviation equal to 3 in., you would expect close to 95% of these males to have heights between 5 ft 8 in. ± 6 in. or between 5 ft 2 in. and 6 ft 2 in.

CHARACTERISTICS OF THE NORMAL CURVE

As shown in Figure 8.2, the shape of the normal curve looks like a bell; it is high in the middle and low at both ends or *tails*. Values in the middle of the curve (close to the mean of the distribution) have a greater probability of occurring than values further away (nearer the tails of the distribution). For example, you should expect z values between 0 and 1 to occur more frequently than z values between 1 and 2; even though the score distances between the mean and $z = +1$ and between $z = +1$ and $z = +2$ are the same, the *area* between the mean and $z = +1$ is much greater than the area between $z = +1$ and $z = +2$. As was observed in Chapter 2, the normal curve is unimodal (has one peak) and symmetrical. (However, not all curves which are bell-shaped, unimodal, and symmetrical are normal—that is, defined by the normal curve equation.)

Notice that the normal curve does not touch the X axis until it reaches ($+$ and $-$) infinity. Extreme positive and negative values can occur, but are extremely unlikely. You *could* actually obtain z values of 5, 6, or more, but such z values would be very rare under this model (and might therefore suggest the possibility of an error in computation).

ILLUSTRATIVE EXAMPLES

The College Boards are administered each year to many thousands of high school seniors and the scores are transformed in such a way as to yield a mean of 500 and a standard deviation of 100 (the same as SAT scores in Chapter 6). These scores are close to being normally distributed, so we can use the normal curve model in our calculations. Since we know both μ and σ, we can convert any College Board score (X) to a z score and refer it to Table B, using

$$z = \frac{X - \mu}{\sigma}$$

$$= \frac{X - 500}{100}$$

(In situations where μ and σ are not known, they can be estimated from the sample $\bar{X}$ and s provided that the sample size is large.)

There are many problems that can be solved by the use of this model:

1. What percent of high school seniors in the population can be expected to have College Board scores between 500 and 675?

The first step in any normal curve problem is to draw a rough diagram and indicate the answer called for by the problem; this will prevent careless errors. The desired percent is expressed as the unshaded area in Figure 8.3.

To obtain the percent area between the mean (500) and a raw score of 675, convert the raw score to a z score:

$$z = \frac{675 - 500}{100} = +1.75$$

Table B reveals that the percent area between the mean and a z score of +1.75 is equal to 45.99%. Therefore, approximately 46% of the population is expected to have scores between 500 and 675 (and thanks to the normal curve model, you can reach this conclusion without having to investigate the entire population).

2. What percent of the population can be expected to have scores between 450 and 500?

Figure 8.3 *Percent of population with college board scores between 500 and 675.*

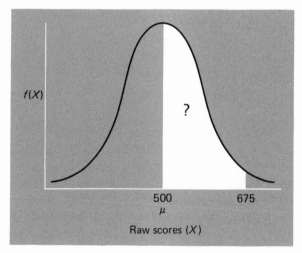

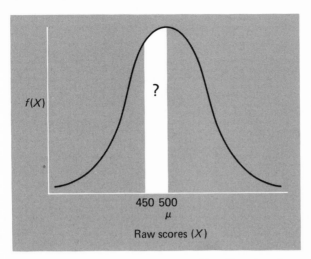

Figure 8.4 *Percent of population with college board scores between 450 and 500.*

The diagram for this problem is shown in Figure 8.4. Converting the raw score of 450 to a *z* score yields:

$$z = \frac{450 - 500}{100} = -0.50$$

Table B does not include any negative *z* values, but entering the table at the *z* score of +.50 will produce the right answer because of the symmetry of the normal curve. The solution is that 19.15%, or about 19% of the population is expected to have scores between 450 and 500.

3. What percent of the population can be expected to have scores between 367 and 540?

This problem is likely to produce some confusion unless you use the diagram in Figure 8.5 as a guide. Table B gives values only between the mean and a given *z* score, so two steps are required. First obtain the percent between 367 and the mean (500); then obtain the percent between the mean and 540; then *add* the two values together to get the desired area. The *z* values are:

$$z = \frac{367 - 500}{100} = -1.33; \quad z = \frac{540 - 500}{100} + 0.40$$

Table B indicates that 40.82% of the total area falls between the mean and a *z* of −1.33, and 15.54% falls between the mean and a *z* of +0.40. Thus, the percent expected to have scores between 367 and 540 is equal to 40.82% + 15.54% or 56.36%.

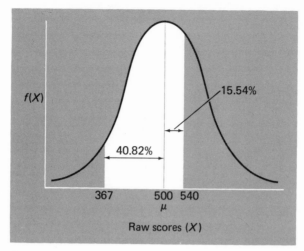

Figure 8.5 *Percent of population with college board scores between 367 and 540.*

4. What percent of the population can be expected to have scores betwen 633 and 700?

Once again, you are likely to encounter some difficulty unless you draw a diagram such as the one shown in Figure 8.6. The z score corresponding to a raw score of 633 is equal to $(633 - 500)/100$ or $+1.33$, and the z score corresponding to a raw score of 700 is equal to $(700 - 500)/100$ or $+2.00$. According to Table B, 40.82% of the curve falls between the mean and $z = +1.33$, and 47.72% of the curve falls between the mean and $z = +2.00$.

Figure 8.6 *Percent of population with college board scores between 633 and 700.*

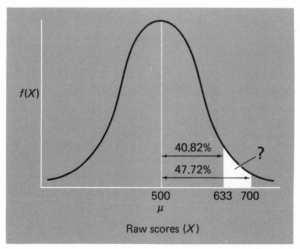

Thus, the answer is equal to 47.72% *minus* 40.82%, or 6.90%. In this example, subtraction (rather than addition) is the route to the correct answer, and drawing the illustrative diagram will help you decide on the proper procedure in each case.

5. What percent of the population can be expected to have scores above 725 ?

The percent to be calculated is shown in Figure 8.7. The z score corresponding to a raw score of 725 is equal to (725 − 500)/100 or +2.25, and the area between the mean and this z score as obtained from Table B is equal to 48.78%. At this point, avoid the incorrect conclusion that the value from the table is the answer

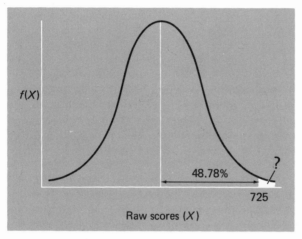

Figure 8.7 *Percent of population with college board scores above 725.*

to the problem. As the diagram shows, you need to find the percent *above* 725, and this can readily be done by noting that the half of the curve to the right of the mean is equal to 50.00%. Thus, the answer is equal to 50.00% − 48.78%, or 1.22%. Note that we have incidentally found that 725 is at about the 99th percentile, since about 1% of the population exceeds it.

6. What is the probability that a person drawn at random from the population will have a score of 725 or more ?

This problem is identical to the preceding one, but is expressed in slightly different terminology. As before, convert 725 to a z score, obtain the percent between the mean and 725, and subtract from 50% to obtain the answer of 1.22%. To express this as a probability, simply convert to a decimal; thus, the probability that any one individual drawn at random from the population will have a score of 725 or more is .0122 (or about .01). Since the probability is approximately 1/100, the *odds against* randomly selecting such a high-scoring person are approximately 99 to 1.

7. A snobbish individual wishes to invite only the top 10% of the population on the college boards to join a club that he is forming. What cutting score should he use to accept and reject candidates?

This problem is the reverse of the ones we have looked at up to this point. Previously, you were given a score and asked to find a percent; this time, the percent is specified and a score value is needed. The problem is illustrated in Figure 8.8.

The needed value is the raw score value corresponding to the cutting line in Figure 8.8. The steps are exactly the reverse of the previous procedure:

(1) If 10% are above the cutting line, then 40% (50% − 10%) are between the mean and the cutting line.

(2) Enter Table B in the *body* of the table (where the percents are given) with the value 40.00 (=40%). Read out the z score corresponding to the percent nearest in value to 40.00 ($z = 1.28$ for 39.97).

(3) Determine the *sign* of the z score. Since the diagram shows that the desired cutting score is *above* the mean, the z score is positive and equals +1.28.

(4) Convert the z score to a raw score. Since $z = (X - \mu)/\sigma$, some simple algebraic manipulations yield $X = z\sigma + \mu$.

Figure 8.8 *Raw score which demarcates the top 10% of the population on college board scores.*

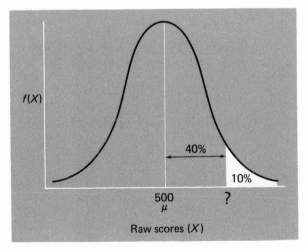

Thus, $X = (+1.28)(100) + 500$ or 628. Thus, any individual with a score of 628 or more may enter the elite club; individuals with scores of 627 or less are summarily rejected.

8. What cutting score separates the top 60% of the population from the bottom 40%?

Since you are given a percent and asked to find a score, this is another "reverse" problem; you need to find the raw score illustrated in Figure 8.9.

The percent with which to enter the table is not immediately obvious, and a hasty reliance on any percent that seems to be handy (such as 60% or 40%) will produce an incorrect answer. The area between the mean and the cutting line is 10%, and it is this value that must be used to enter Table B. As in the previous problem, enter the table in the center; find the value nearest to 10.00 (here, 09.87) and read out the corresponding z score (0.25). Next, determine the sign; since the diagram shows that the desired cutting score is *below* the mean, the sign is negative and $z = -0.25$. Finally, convert this value to a raw score; $X = (-0.25)(100) + (500)$ or 475. Thus, the cutting score between the top 60% and the bottom 40% of the population is expected to be a score of 475.

As you can see, the normal curve is useful for solving various types of problems involving scores and percentage distributions. Of even greater importance is its use with certain sampling distributions, such as means, to which we now turn.

Figure 8.9 *Raw score demarcating the top 60% of the population on college board scores.*

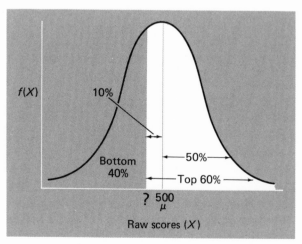

SUMMARY

One frequently used theoretical model is the *normal curve*. One use of this model is as follows: If it is assumed that raw scores in the population are normally distributed, and that we know or can estimate the population mean and standard deviation, one can obtain (1) the percent of the population with raw scores above or below or between specified values and/or (2) the raw score which demarcates specified percents in the population.

9 inferences about the mean of a single population

In the preceding chapter, a crucial assumption was made—that the *raw scores* in the population were normally distributed. This assumption enabled you to draw inferences about the percent of cases in the population between specified raw score values (for example, that about 46% of the people in the population were expected to have college board scores between 500 and 675) without having to measure the entire population.

The present chapter deals with a second, more important use of the normal curve model: drawing inferences based on a *statistic* whose *sampling distribution* is normal. While the sampling distributions of several statistics are of normal form, we will use only the distribution of sample *means* to illustrate the use of the normal curve model. Our objective will be to draw inferences about the mean of a population (μ); for other statistics the inferential strategy remains the same, but the details of the procedures are different.

THE STANDARD ERROR OF THE MEAN

Suppose that you are interested in studies of heights of American men, a variable known to be approximately normally distributed in the population. If many, many random samples of the same size were selected from this population and the mean of each sample were computed, the *sample means* when plotted as a frequency polygon would be normally distributed about *their* mean, which is also the population mean. As a matter of fact, even if the distribution of raw scores in the population is *not* normal, the sample means would still be approximately normally distributed *if* the number of cases in each sample is sufficiently large. Thus, the normal curve model can be used to represent the sampling distribution of the statistic $\bar{X}$.

Before you can proceed to draw inferences about the population mean, however, you need a measure of the variability of this sampling distribution. The standard deviation of the raw scores σ (or its estimate s) is *not* the correct measure to use. To see why, consider once again the normal distribution of college board scores in the population ($\mu = 500$). Selecting one man at random with a score of 725 is fairly unlikely, but drawing *a whole sample of people* with scores so extreme that the sample averages out to a mean of 725 is much less likely. In other words, an extreme value of the *sample mean* is less likely than is an extreme value of a single raw score, since a greater number of low probability events must occur simultaneously for the sample mean to be extremely different from the population mean. This implies that the variability of the distribution of sample means is *less* than the variability of the distribution of raw scores; it decreases as the sample N increases.

If a large number of sample means were actually available, you could measure the variability of the means directly by calculating the standard deviation *of the sample means*, using the usual formula for computing the standard deviation of a set of numbers (Chapter 5) and treating the sample means just like ordinary numbers. However, the behavioral scientist never has *many* samples of a given size; he usually has only one.

Fortunately, as a result of some mathematical–statistical theory, the variability of the distribution of sample means can be *estimated* from a *single* random sample drawn from the population. It is important that the sample be randomly drawn; if you should happen to be restricted to atypical samples such as basketball players with regard to height, your attempt to make reliable estimates of parameters in the general population of males would not meet with outstanding success.

Although the proof is beyond the scope of this book, it can be shown that the *estimated* standard deviation of the sampling distribution of means of samples of size N, called the *standard error of the mean* and symbolized by $s_{\bar{X}}$, is given by:

$$s_{\bar{X}} = \frac{s}{\sqrt{N}}$$

where s = standard deviation of the sample raw scores
N = number of observations in the sample

This quantity estimates the variability of means for samples of the given N, so it is a *standard deviation* of *means*. But since it does that, it also tells you how trustworthy is the single mean which you have in hand, hence the name *standard error* of the mean.

As you can see from the formula, the standard error of the mean must be less than the standard deviation of the raw scores, and it becomes smaller as the sample size grows larger, just as would be expected from the preceding discussion. The term *standard error* is used to express the idea that the difference between a population mean and the mean of a sample drawn from that population is an "error" caused by sampling (that is, the cases that happened by chance to be included in the sample). If *no* error existed (as would happen if the "sample" consisted of the entire population and $N = \infty$), all sample means would exactly equal each other and the value of the standard error of the mean would be zero. This situation would be ideal for purposes of statistical inference, since the mean of any one sample would exactly equal the population mean; sad to say, it never happens—actual data are always subject to sampling error.

To clarify the preceding discussion, let us assume that the mean height in the population of American men is equal to 68 in., and compare the distribution of raw scores to the distribution of sample means. A hypothetical distribution of five million height observations for American men is shown in Figure 9.1; each X represents

Figure 9.1 *Distribution of observations of heights (in.) for the population of 5,000,000 adult males.*

$X_1 = 68.3$ in. $X_2 = 66.2$ in. $X_3 = 73.1$ in.   ... $X_{5,000,000} = 69.3$ in.

Observation 1   *Observation 2*   *Observation 3*   *Observation 5,000,000*

or   *or*   *or*   *or*

Individual 1   *Individual 2*   *Individual 3*   *Individual 5,000,000*

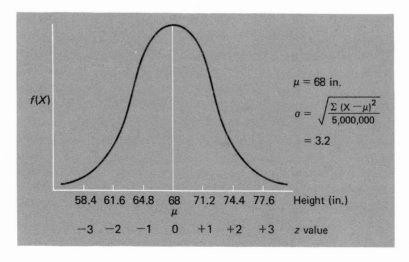

one *raw score*. Now suppose 1000 random samples of $N = 100$ observations each are randomly selected from the population and the mean of each sample is calculated. The resulting distribution of *sample means* is illustrated in Figure 9.2.

Notice that the shape of *both* distributions (original distribution of raw scores or single observations, and distribution of sample means) is normal and centered about μ (68 in.). However, the distribution of sample means is *less variable* than is the distribution of raw scores. The standard deviation of the raw scores shown in Figure 9.1 would be calculated directly by applying the usual formula for the standard deviation to the 5,000,000 raw scores; let us suppose that σ is found to be equal to 3.2 in. The standard deviation of the sample means in Figure 9.2 would be calculated directly by applying the usual formula for the standard deviation (Chapter 5) to the 1000 sample means; as the diagram indicates, this is equal to only 0.32 in.*

Figure 9.2 *Distribution of mean heights (in.) for 1000 randomly selected samples of N = 100 height observations in each sample.*

$\bar{X}_1 = 68.1$ in. $\bar{X}_2 = 67.8$ in. $\bar{X}_3 = 68.4$ in. ... $\bar{X}_{1000} = 67.7$ in.

Sample 1   *Sample 2*   *Sample 3*   *Sample 1000*

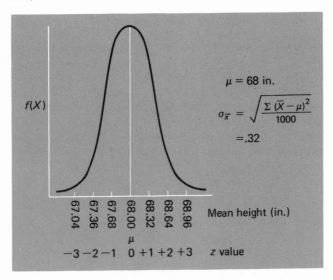

* When the population standard deviation is known, the standard error of the mean is symbolized by $\sigma_{\bar{x}}$ and is found from $\sigma_{\bar{x}} = \sigma/\sqrt{N}$, where $\sigma =$ the standard deviation of the population raw scores and $N =$ the number of observations in the random sample. Thus, in the present case, the value of $\sigma_{\bar{x}}$ would be $3.2/\sqrt{100}$ or .32.

Although this highly unrealistic situation is useful for purposes of illustrating the standard error of the mean, let us return to the real world wherein you will have only one sample at your disposal. Suppose that you randomly draw *one* sample of $N = 100$ *midwestern* American men, and you calculate the mean height of this sample as $\bar{X} = 67.4$ in. and the standard deviation of the raw scores as $s = 3.5$ in. Then, the standard error of the mean can be *estimated* as follows:

$$s_{\bar{x}} = \frac{s}{\sqrt{N}} = \frac{3.5}{\sqrt{100}} = .35$$

You would now like to know whether the average height of midwestern men is different from the average of the entire U.S. adult male population (68 in.). Without measuring the entire population of midwestern men, you cannot answer this question with certainty, but you can use the sample mean and the standard error of the mean to draw an inference about the population which will indicate the answer warranted by the sample data. There are two possibilities: *either* the sample with a mean of 67.4 does come from a population with a mean of 68, and the difference of .6 between these two values is due to sampling error (the cases that happened to fall in the sample of 100 midwestern men that you selected at random had a mean which fell .6 below the population mean), *or* the sample did *not* come from a population with a mean of 68 but from one whose mean was some other value. In order to decide between these alternatives, it would be very helpful to know how *unusual* such a sample would be—that is, the probability of selecting a sample with $\bar{X} = 67.4$ *if* $\mu = 68$. Which of the two alternatives is more justified by the experimental data?

HYPOTHESIS TESTING

In order to answer this question, it is necessary to *test the hypothesis* that the mean of the population from which the sample of 100 midwestern men was drawn is equal to 68 in. The testing of a statistical hypothesis is designed to help in making a decision about a population parameter (here, μ). In Chapter 7, we saw that if a series of experiments is run in each of which a coin is flipped six times, six heads out of six flips should be obtained approximately once or twice in 100 experiments *if* the coin is fair, that is, under the hypothesis $P(H) = 1/2$. Since this was "sufficiently unlikely" in terms of the ".05 decision rule," the hypothesis that the coin was fair (H_0) was *rejected* in favor of the hypothesis that the coin was biased (H_1). The procedure is similar in the present height study; it is necessary to determine the probability of drawing a sample with a mean of 67.4 in. *if* the population mean is in fact 68 in. and compare this probability to a decision rule.

THE NULL AND ALTERNATIVE HYPOTHESES

The first step is to state the *null hypothesis* (symbolized by H_0), which specifies the hypothesized population parameter. In the present study,

$$H_0: \quad \mu = 68$$

This null hypothesis implies that the sample, with mean equal to 67.4 in., is a random sample from the population with μ equal to 68 in. (and that the difference betweeen 67.4 and 68 is due to sampling error). As in the coin experiment, the probability of obtaining a sample mean of the observed value (67.4) is calculated under the assumption that H_0 is true (that is, that $\mu = 68$).

Next, an *alternative hypothesis* (symbolized by H_1) is formed; it states another value or set of values for the population parameter. In the present study,

$$H_1: \quad \mu \neq 68$$

This alternative hypothesis states that the population from which the sample comes has *not* a μ equal to 68 in.; that is, the difference of .6 between the sample mean and the null-hypothesized population mean is due to the fact that the null hypothesis is incorrect.

CONSEQUENCES OF POSSIBLE DECISIONS

If the probability of obtaining a sample mean of 67.4 when $\mu = 68$ is deemed "sufficiently unlikely" to occur by a decision rule such as the ".05 rule," H_0 is rejected in favor of H_1. If the probability that is calculated is "not sufficiently unlikely," H_0 is retained. While the null is *hypothesized* to be true at the outset, the statistical test *never proves* whether the null hypothesis is true or false. The test merely indicates whether or not H_0 is "sufficiently unlikely" given the decision rule. Without measuring the entire population, it is impossible to *prove* anything, and the final decision that you reach could be correct or incorrect. In fact, there are four possible eventualities*:

---

* Compare this discussion with that of flipping coins with a friend in Chapter 7.

1. The population mean is actually 68 (that is, the sample comes from a population where $\mu = 68$) and you *incorrectly reject* H_0. Rejecting a *true* null hypothesis is called a *Type I Error*, and the probability (or *risk*) of committing such an error is symbolized by the Greek letter alpha (α).

2. The population mean is actually *not* 68 and you *incorrectly retain* H_0. Failing to reject a *false* null hypothesis is called a *Type II Error*, and the probability (risk) of committing this *faux pas* is symbolized by the Greek letter beta (β).

3. The population mean is actually *not* 68 and you *correctly reject* H_0. The probability of reaching this correct decision is called the *power* of the statistical test and is equal to $1 - \beta$.

4. The population mean is actually 68 and you *correctly retain* H_0. The probability of making this correct decision is equal to $1 - \alpha$.

The four eventualities are summarized in Table 9.1.

Table 9.1   *Model for error risks in hypothesis testing*

	state of the population	
	H_0 *is actually true*	H_0 *is actually false*
outcome of experiment dictates: **retain** H_0	correct decision: probability of retaining true H_0 is $1 - \alpha$	type II error: probability (risk) of retaining false H_0 is β
reject H_0	type I error: probability (risk) of rejecting true H_0 is α	correct decision: probability of rejecting false H_0 (*power*) is $1 - \beta$

Whenever you conduct a statistical test, there are always two possible kinds of error risk: rejecting a null hypothesis which is really true (alpha risk) and failing to reject a null hypothesis which is really false (beta risk). An important aspect of such statistical tests is that it is possible to specify and control the probability of making each of the two kinds of errors. For the present, we will be concerned primarily with Type I error and the corresponding alpha risk; Type II error and beta risk will be treated in Chapter 13.

The reason why these errors cannot be summarily dismissed with an instruction to make the probability of their occurrence as small as possible is that for a sample of a given size, making a Type I error *less* likely simultaneously makes a Type II error *more* likely, and vice versa. The task is therefore the nontrivial one of balancing the desirable aspects of reducing the probability of one kind of error against the undesirable aspects of increasing the probability of the other kind of error. This is a problem of real practical concern to the behavioral scientist. Usually, the theory which the scientist hopes to support is identified with the *alternative* hypothesis; thus, rejecting H_0 will cause the scientist to conclude that his theory has been supported and failing to reject H_0 will cause him to conclude that his theory has not been supported. This is a sensible procedure, since proper scientific caution dictates that the researcher should not jump to conclusions; he should claim success only when there is a strong indication to that effect—that is, when the results are sufficiently striking to cause rejection of H_0. If the researcher commits a Type I error (rejects H_0 when it is in fact true), he will triumphantly conclude that his theory has been supported when in fact the theory is incorrect. This will cause him a professional "black eye" when other researchers attempt to replicate his findings and are unable to do so, an eventuality similar to the coin experiment wherein a Type I error—concluding that the coin is biased when it is in fact fair—might well result in a physical "black eye" from an honest but outraged friend. Thus, the researcher might be tempted to make the probability of a Type I error as small as possible. Unfortunately, doing so will increase the chances of a Type II error, which also has rather drastic consequences. If the researcher fails to reject H_0 when it is in fact false, he will erroneously fail to conclude that his theory is valid. Thus, he will miss out on the chance to advance the state of scientific knowledge, and will not receive the recognition due him for his correct theory. This problem is resolved in part by the "decision rule" that is selected.

THE CRITERION OF SIGNIFICANCE

Having set up the null and alternative hypotheses, a rule is needed indicating *how unusual* a sample mean must be to cause you to reject H_0 and accept H_1. Intuitively, an observed sample mean of 68.1 would hardly represent persuasive evidence against the null hypothesis that $\mu = 68$, even though it is not exactly equal to the hypothesized population value; while an observed sample mean of 40 would do more to suggest that the hypothesized population value was incorrect. Where should the line be drawn? As was pointed out in Chapter 7, there are no *absolute* rules for establishing values at which a sample mean is considered to be so deviant that H_0 *must* be rejected. All that the statistical test can do is to compare hypotheses about a population parameter and yield a statement about the probability of an experimental result *if H_0 is true*.

The numerical value specified by the decision rule that you select is called the *criterion of significance*; thus, when you choose the ".05 decision rule," you are using the 5% or *.05 criterion of significance*. Since the criterion of significance is denoted by α, the .05 criterion can be expressed in symbols as $\alpha = .05$. As was indicated previously, the value of α specifies the probability of making a Type I error; thus, when you use the .05 criterion of significance, the probability of falsely concluding that $\mu \neq 68$ is equal to .05. This means that if you were to repeat this experiment many times when the null hypothesis is true, an average of one out of every 20 decisions would be to (incorrectly) reject the null hypothesis when you use the .05 criterion. [Note, however, that the probability of falsely retaining the null hypothesis that $\mu = 68$, a Type II error, is not so conveniently determined (see Chapter 13).]

You should reject H_0 if the experimental result is unusually deviant (high or low). This is called a *"two-tailed" test of the null hypothesis* using the .05 criterion of significance; it is illustrated in Figure 9.3. Note that the circumstances portrayed are for the null hypothesis to be *true*, and the probability of falsely concluding that $\mu \neq 68$ is equal to $2\frac{1}{2} + 2\frac{1}{2}\% = .05$. When the normal curve

Figure 9.3 *Areas of rejecting H_0 in both tails of the normal sampling distribution using the .05 criterion of significance when H_0 actually is true.*

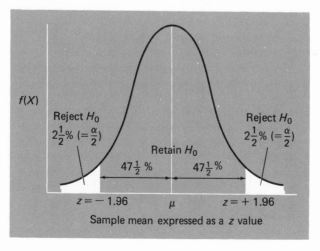

model is appropriate, the cutoff scores beyond which H_0 should be rejected are easily ascertained using the normal curve table (Table B). Since 2.5% of the area of the curve is in each tail, $50.0 - 2.5\%$ or 47.5% falls between the mean and the cutoff score. Entering Table B in the center with a value of 47.5%, the cutting scores expressed as z values are found to be equal to -1.96 and $+1.96$. Thus, once the sample mean has been expressed as a z value, the .05 criterion of significance states that H_0 should be rejected if z is less than or equal to -1.96 or z is greater than or equal to $+1.96$, and H_0 should be retained if z is between -1.96 and $+1.96$. Note that the area in each rejection region is equal to $a/2$ $(.05/2 = .025)$, and that the total *area of rejection* (or *critical region*) is equal to the significance criterion a or .05. Thus, H_0 will be retained if the probability of obtaining the observed sample mean *if H_0 is true* is greater than .05; and H_0 will be rejected if the probability is no larger than .05.

Another traditional, but equally arbitrary, criterion is the 1% or *.01 criterion of significance*. Using this criterion, two-tailed hypotheses about the population mean are rejected if, when H_0 is true, a sample mean is so unlikely to occur that no more than 1% of sample means would be so extreme, with .005, or 1/2 of 1%, in each tail. This criterion is illustrated in Figure 9.4.

When the normal curve model is appropriate, the cutoff scores beyond which H_0 should be rejected using the .01 criterion are

Figure 9.4   *Areas of rejecting H_0 in both tails of the normal sampling distribution using the .01 criterion of significance when H_0 actually is true.*

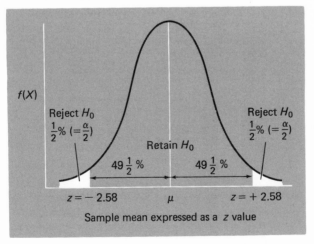

ascertained by entering Table B in the body of the table with the value 49.5%; the cutting scores expressed as z values are -2.58 and $+2.58$. Thus, once the experimental result (sample mean) has been expressed as a z value, the .01 criterion of significance states that H_0 should be rejected if z is less than or equal to -2.58 or z is greater than or equal to $+2.58$, and H_0 should be retained if z is between -2.58 and $+2.58$. Once again, the area of rejection is equal to a $(0.5 + 0.5\% = 1\%$ or $.01)$, the criterion of significance that is being used.

The 1% criterion of significance is a more stringent test for rejecting H_0 than is the 5% criterion, since only *very* unusual sample means will be considered "sufficiently unlikely." It has the corresponding advantage of making the probability of a Type I error smaller. However, the other side of the coin is that the 1% criterion is more lenient about retaining H_0, and consequently the probability of a Type II error is greater. When you use the 1% criterion, you will reject H_0 only in more extreme situations (when $P \leq .01$) and are therefore more likely to be correct (and avoid a Type I error) when you do reject H_0; but you must as a consequence fail to reject H_0 in some rather unlikely situations (that is, whenever $P > .01$, even if it is less than .05) and are more likely to be wrong when you do retain H_0. (Similarly, even more stringent criteria of significance, such as the .001 criterion, would further reduce the chances of a Type I error at the cost of further increasing the chances of a Type II error.)

The most frequently used criterion of significance in behavioral research is the .05 criterion, which keeps the chances of a Type II error (missing out on a new finding) lower than for the .01 criterion. It is generally felt that this criterion is sufficiently conservative in terms of avoiding a Type I error (the probability of which equals .05). (A .10 criterion of significance would reduce the chances of a Type II error, but is usually regarded as too likely to lead to a Type I error to be acceptable.) Keep in mind, however, that the value of .05 is a highly arbitrary choice, and one should not be rigid about values very close to the cutting line. If for example $z = 1.88$, the decision rule requires you to retain H_0. It would be a shocking violation of sound research practice (and ethics) to slide the cutting score down a few notches in order to reject H_0 and support your theory, but it is admittedly frustrating to fall just short of the critical value of $+1.96$. The sensible thing to do in such a close case is to repeat the experiment if possible (so as to get more data on which to base a decision).

THE STATISTICAL TEST FOR THE MEAN OF A SINGLE POPULATION WHEN σ IS KNOWN

Let us suppose that you select the .05 criterion of significance in the height experiment. Also assume that you know that the *population* standard deviation, σ, is equal to 3.2. (The statistical test used when σ is *not* known is discussed in the next section.) Then, the sample mean must be expressed as a z value in order to compare it to the criterion values of ±1.96. However, since you are drawing inferences about the *mean* of a single population, the *standard error of the mean* must be used as the measure of variability. That is,

$$z = \frac{\bar{X} - \mu}{\sigma_{\bar{x}}}$$

where $\bar{X}$ = the observed value of the sample mean
μ = the hypothesized value of the population mean
$\sigma_{\bar{x}}$ = the estimated standard error of the mean when σ is known $(= \sigma/\sqrt{N})$

Given that σ = 3.2,

$$\sigma_{\bar{x}} = \frac{3.2}{\sqrt{100}} = .32$$

$$z = \frac{67.4 - 68}{.32}$$

$$= \frac{-.60}{.32}$$

$$= -1.875$$

The z value of −1.875 falls between the critical values of −1.96 and +1.96 and hence is *not* in the area of rejection; consequently, your decision should be to retain H_0. Since the probability of a Type II error (failing to reject H_0 when H_0 is actually false) is not conveniently determined, however, you do not know how confident to be in your decision to retain H_0. You should therefore conclude only that *there is not sufficient reason to reject the null hypothesis* that the average height for midwestern American men is equal to the average height for the typical man. (The determination of the probability of a Type II error is discussed in Chapter 13.) In other words, a sample mean of 67.4 is " not sufficiently unlikely" to occur if $\mu = 68$ for you to reject H_0. Since −1.875 is very close to −1.96, you might well decide to repeat the experiment with a new sample in order to gain additional information.

Now let us suppose instead that it is known that $\sigma = 3.0$. In this situation,

$$\sigma_{\bar{x}} = \frac{3.0}{\sqrt{100}} = .30$$

$$z = \frac{67.4 - 68}{.30}$$

$$= -2.00$$

This z value of −2.00 falls beyond the critical value of −1.96 and hence *is* in the area of rejection ; consequently, your decision in this instance should be to reject H_0 in favor of H_1 and conclude that the average height for midwestern American men is *not* equal to (in fact, is less than) the average height for the typical American man. (Midwestern readers please note that this is an entirely hypothetical result.) The statistical test has indicated that a sample mean of 67.4 is "sufficiently unlikely" to occur if $\mu = 68$ for you to reject H_0. That is, according to the model, if H_0 were true and many, many random samples of size 100 were drawn from the population, you would expect to get a sample mean as or more deviant from 68 than the one actually obtained less than 5% of the time. Consequently, it is a better bet to abandon the assumption that H_0 is true (and that $\mu = 68$) and prefer the alternative hypothesis. This result is said to be "*statistically significant at the .05 level of confidence*"; in other words, H_0 is rejected at the .05 significance level. (If H_0 is retained, the result is said to be "*not statistically significant at the .05 level.*")

THE STATISTICAL TEST FOR THE MEAN OF A SINGLE POPULATION WHEN σ IS *NOT* KNOWN: THE *t* DISTRIBUTIONS

We have seen that when you want to draw an inference about the mean of a single population, your first task is to set up two hypotheses. The *null hypothesis* (H_0) is a statement about the expected value of the population mean, and the *alternative hypothesis* (H_1) is a statement which includes all other values of the population mean. *If σ is known*, you then estimate the variability of the sampling distribution of means by computing $\sigma_{\bar{x}}$. Then, using $\sigma_{\bar{x}}$ and the value of the population mean specified by H_0, you transform your sample mean to a z value. Since values obtained from the formula $(\bar{X} - \mu)/\sigma_{\bar{X}}$ are normally distributed, the normal curve table enables you to convert the z value to the *probability* of obtaining a sample mean as or more extreme than the one you obtained *if H_0 is true*.

This probability enables you to make a decision, guided by your preselected criterion of significance, about retaining or rejecting H_0. Thus, using the .05 criterion, you *retain* (do *not* reject) H_0 if this probability is greater than .05 (that is, if z is between -1.96 and $+1.96$); your result is "not sufficiently unlikely" to occur if H_0 is true. If, however, this probability is no greater than .05 (that is, if z is -1.96 or less, or $+1.96$ or more), you *reject H_0* in favor of H_1; your result is "sufficiently unlikely" to occur if H_0 is true.

Only rarely, however, is σ known exactly in behavioral science research. Most of the time, it is necessary to compute the estimated standard error of the mean based on the *sample* standard deviation, s, where $s_{\bar{X}} = s/\sqrt{N}$. Values obtained from the formula $(\bar{X} - \mu)/s_{\bar{X}}$ are *not* normally distributed (although the normal curve will yield a good approximation if the sample size is greater than 25 or 30). This is because $s_{\bar{X}}$ is only an estimate of $\sigma_{\bar{X}}$. As a result, using the normal curve procedure will give wrong answers, primarily when the sample size is small, because the "frame of reference" or model is incorrect. Therefore, you must refer to the exact distribution of $(\bar{X} - \mu)/s_{\bar{X}}$, which is appropriate for all sample sizes but particularly necessary for small samples, namely *t distributions*. Your task is somewhat complicated by the fact that there is a *different* t distribution for every sample size. Thus, the sampling distribution of means based on a sample size of 10 has one t distribution, while the sampling distribution of means based on a sample size of 15 has a different t distribution.

Fortunately, you do not have to know the shape of each of the various t distributions; probability tables for the curves corresponding to each sample size have already been determined by mathematicians. All you need do is note the size of your sample and then refer to the proper t distribution. Such a t table is presented as Table C in the Appendix.

DEGREES OF FREEDOM

The t table is given in terms of *degrees of freedom* (df) rather than sample size. When you have a sample mean and a hypothesized population mean (and σ is not known),

$$df = N - 1$$

That is, degrees of freedom in the present situation are equal to *one less* than the sample size, equal, you will recall, to the divisor of s (see Chapter 5). " Degrees of freedom " is a concept which arises at several points in statistical inference; you will encounter it again later in connection with chi square (Chapter 16). The number of degrees of freedom is the number of freely varying quantities in the kind of repeated random sampling which produces sampling distributions. In connection with the test of the mean of a single population, it refers to the fact that each time a sample is drawn and the population variance is estimated, it is based on only $N-1$ df; that is, $N-1$ freely varying quantities. When one finds for each of the N observations its deviation from the sample mean, $X-\bar{X}$, not all these N quantities are free to vary in any sample. Since they must sum to zero (as was shown in Chapter 4), the Nth value is automatically determined once any $N-1$ of them are fixed at given observed values. For example, if the sum of $N-1$ deviations from the mean is -3, the Nth must equal $+3$ so that the sum of all deviations from the mean will equal zero. Thus, in this situation, there are only $N-1$ df, and therefore the t distribution is here based on $N-1$ df. The df and not the N is the fundamental consideration, because df need not be $N-1$ in other applications of t.

THE STATISTICAL TEST

A section of the t table is shown below:

df	...	$t_{.05}$	$t_{.01}$	...
⋮				
4		2.78	4.60	
⋮				
10		2.23	3.17	
11		2.20	3.11	
12		2.18	3.06	
13		2.16	3.01	
14		2.15	2.98	
15		2.13	2.95	
⋮				
40		2.02	2.70	
⋮				
120		1.98	2.62	
⋮				
∞		1.96	2.58	

Each horizontal row actually represents a separate distribution which corresponds to the *df*; the row for ∞ represents the normal distribution. To illustrate the use of the *t* table, suppose that you wish to test the null hypothesis H_0: $\mu = 17$ against the alternative H_1: $\mu \neq 17$ using the .05 criterion of significance. You obtain a sample of $N = 15$ and find that $\bar{X} = 18.80$ and $s = 3.50$. Then compute

$$t = \frac{\bar{X} - \mu}{s_{\bar{x}}}$$

where $\bar{X}$ = the observed value of the sample mean
μ = the hypothesized value of the population mean
$s_{\bar{x}}$ = the estimated standard error of the mean ($= s/\sqrt{N}$)

In the present example,

$$t = \frac{18.80 - 17}{(3.50/\sqrt{15})} = \frac{1.80}{.90} = 2.00$$

You now compare the *t* value that you have computed to the critical value obtained from the *t* table. For 14 degrees of freedom (one less than your *N* of 15) and using the .05 criterion of significance, the critical value is 2.15. A *t* value larger than this *in absolute value* (that is, ignoring the sign) is "sufficiently unlikely" to occur if H_0 is true for you to reject H_0. That is, the probability of obtaining *t* values absolutely larger than 2.15 when $df = 14$ is .05. Therefore, if the absolute value of the *t* that you compute is *less than* 2.15, *retain* H_0. If, however, your result is *greater* in absolute value than 2.15, *reject* H_0 in favor of H_1. (In the unlikely event that your computed *t* value exactly equals the *t* value obtained from the table, reject H_0.) In our example, the obtained *t* value of 2.00 is not larger in absolute value than 2.15; therefore you retain H_0 and conclude that the data do *not* warrant concluding that the population mean is not 17.

If instead your sample size were 12, you would have 11 degrees of freedom and the corresponding critical value from the table (again using the .05 criterion of significance) would be 2.20. You would retain H_0 if the absolute size of the *t* value you computed were less than 2.20, and you would reject H_0 in favor of H_1 if this value were equal to or greater in absolute value than 2.20. Similarly, for the same sample size, the critical value of *t* using the .01 criterion of significance is 3.11; you would retain H_0 if your computed *t* value were no greater than 3.11 in absolute value and would reject H_0 otherwise.

THE *t* AND *z* DISTRIBUTIONS COMPARED

The *t* distribution for four degrees of freedom and the normal curve are compared in Figure 9.5. Note the longer tails in the case of the *t* distribution, indicating that extreme values are more likely to occur by chance than in the case of the normal curve. Thus, while 95% of the normal curve lies between the *z* values −1.96 and +1.96, 95% of the *t* distribution for four degrees of freedom lies between −2.78 and +2.78. Therefore, using the .05 criterion, 2.78 (and *not* 1.96) must be used as the critical value when σ is not known and *df* = 4.

Notice also in the *t* table that the *smaller* the sample size, the *larger* *t* must be in order to reject H_0 at the same criterion of significance. Also, when the sample size is large, *z* and *t* are virtually equivalent. Thus, for 40 degrees of freedom, *t* = 2.02 (using the .05 criterion), while for 120 degrees of freedom, *t* = 1.98. These are so close to the *z* value of 1.96 that it makes little difference which one you use. For example, suppose that in the study of heights of midwestern American men discussed earlier in this chapter, a sample of 100 midwestern men is found to have $\bar{X} = 67.4$ in. and $s = 3.5$ in. and σ is *not* known. To test the null hypothesis that μ = 68, you would compute

$$s_{\bar{x}} = \frac{3.5}{\sqrt{100}} = .35$$

$$t = \frac{67.4 - 68}{.35}$$

$$= -1.71$$

Figure 9.5   *Comparison between the normal curve and the t distribution for df = 4.*

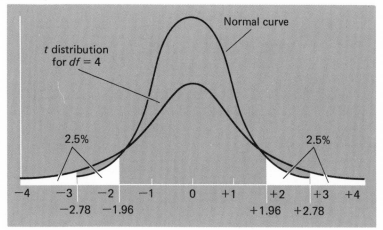

You would decide to retain H_0 whether you compared 1.71 (the absolute value) to the critical t value of 1.99 for $df = 99$ or to the z value of 1.96. In behavioral science research, it is customary to report the obtained value as a t value even for large samples, since although z is a good approximation in large samples, it is still an approximation. Thus, you should use the critical t value obtained from the t table (rather than the z value of 1.96 for the .05 criterion or 2.58 for the .01 criterion) unless you happen to know the exact value of σ. *Retain H_0* if your computed t value (ignoring the sign) is smaller than the critical value, and *reject H_0* otherwise.

INTERVAL ESTIMATION

Another way of solving the problem of whether to retain or reject H_0 is to estimate an *interval* within which the population mean is likely to fall. If the value of μ assumed under the null hypothesis does *not* fall in the interval, H_0 is rejected. The interval represents an estimate of how far away from the sample mean the population mean is likely to be; that is, how much samples are likely to vary from the population mean due to sampling error. Such an interval is called a *confidence interval* because you can be reasonably confident that the population mean falls within that interval; the end points of the interval are called the *confidence limits.* Confidence intervals have the advantage of specifying all values of μ for which H_0 should be retained, which permits several different hypothesized values of μ to be tested without involving any extra computational effort.

For example, suppose that in the height experiment discussed earlier in this chapter, you would like to know whether the sample is or is not unlikely to have come from each of the following five populations:

A $\mu = 67.7$ in.
B $\mu = 68$ in.
C $\mu = 68.2$ in.
D $\mu = 67.1$ in.
E $\mu = 66.6$ in.

Let us first tackle the problem the long way and test each hypothesized value of the population mean separately, assuming σ is not known and using the previously obtained values of $\bar{X} = 67.4$ in. and $s_{\bar{x}} = .35$ in. Under null hypothesis A, the observed sample mean of 67.4 in. would have a t value of

$$t = \frac{67.4 - 67.7}{.35} = -.86$$

Since the distance between the observed value of the sample mean and the hypothesized value of the population mean is less in absolute value than the critical value of t for $df = 99$ of 1.99, you would retain H_0 and conclude that there is not sufficient reason to reject the hypothesis that the population mean is 67.7. The graphic representation is shown in Figure 9.6A.

Figure 9.6 *Distance between the observed sample mean and the population mean under varying hypotheses about μ (σ is not known).*

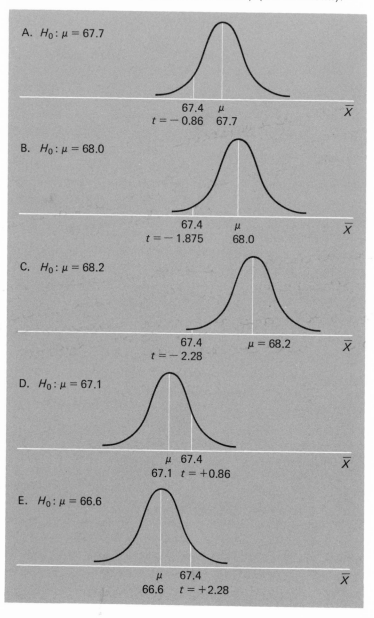

A. $H_0: \mu = 67.7$

       67.4 μ
    $t = -0.86$ 67.7 $\overline{X}$

B. $H_0: \mu = 68.0$

       67.4 μ
    $t = -1.875$ 68.0 $\overline{X}$

C. $H_0: \mu = 68.2$

       67.4 $\mu = 68.2$
    $t = -2.28$ $\overline{X}$

D. $H_0: \mu = 67.1$

     μ 67.4
    67.1 $t = +0.86$ $\overline{X}$

E. $H_0: \mu = 66.6$

     μ 67.4
    66.6 $t = +2.28$ $\overline{X}$

If μ were posited as 68 (null hypothesis B), would 67.4 in. be an unusual sample mean? (See Figure 9.6B.) As we have already seen, $t = -1.875$, which is not sufficiently unusual according to the .05 criterion since more deviant results would be expected to occur *more* than 5% of the time if $μ = 68$. Consequently, the null hypothesis that $μ = 68$ is *not* rejected.

If μ were assumed equal to 68.2 in. (null hypothesis C), a sample mean of 67.4 *would* be unusual:

$$t = \frac{67.4 - 68.2}{.35} = -2.28$$

The absolute value of the obtained t is *greater* than the critical value of 1.99 for $df = 99$. Therefore, this null hypothesis is rejected; a population mean of 68.2 in. is sufficiently unlikely for you to reject it. (See Figure 9.6C.)

Figures 9.6D and 9.6E illustrate examples where the sample mean is larger than the hypothesized population mean. In Figure 9.6D, the sample mean of 67.4 in. is *not* unusually deviant from the hypothesized population mean of 67.1, because the t value of $+0.86$ is less than the critical value of 1.99. In Figure 9.6E, however, where μ is assumed to be 66.6 in., the sample mean of 67.4 in. *is* an unusual result ($t = +2.28$), so you would reject H_0 and conclude that a population mean of 66.6 in. is *not* likely.

You will probably agree that computing a new t value for every hypothesized value of μ is not very efficient. A better plan is to set up an interval to include all "not sufficiently unlikely" values of μ (all values for which H_0 should be retained). Using the .05 criterion of significance and $df = 99$, the furthest that any sample mean can lie from μ and still be retained as "not sufficiently unlikely" is $\pm 1.99 s_{\bar{x}}$ (where 1.99 is the critical value of t for the specified df, 99). That is, the upper limit of the confidence interval may be found by solving the equation

$$+1.99 = \frac{\bar{X} - μ}{s_{\bar{x}}}$$

for μ and the lower limit of the confidence interval may be found by solving the equation

$$-1.99 = \frac{\bar{X} - μ}{s_{\bar{x}}}$$

for μ. This is readily done as follows:

$$\pm 1.99 s_{\bar{x}} = \bar{X} - μ$$
$$μ = \bar{X} \pm 1.99 s_{\bar{x}}$$

or

$$\bar{X} - 1.99s_{\bar{x}} \leq \mu \leq \bar{X} + 1.99s_{\bar{x}}$$

Thus,

$$67.4 - 1.99(.35) \leq \mu \leq 67.4 + 1.99(.35)$$
$$67.4 - .70 \leq \mu \leq 67.4 + .70$$
$$66.7 \leq \mu \leq 68.1$$

Thus, you should retain any hypothesized value of μ between 66.7 in. and 68.1 in., inclusive, using the confidence interval corresponding to the .05 criterion of significance (called the *95% confidence interval*) and reject any hypothesized value of μ outside this interval.

The general formula for computing confidence intervals when σ is not known is:

$$\bar{X} - ts_{\bar{x}} \leq \mu \leq \bar{X} + ts_{\bar{x}}$$

where $t =$ critical value from t table for $df = N - 1$
   $\bar{X} =$ observed value of the sample mean

The top part of Figure 9.7 illustrates the limits of the 95% confidence interval. Notice that the sample, with mean $\bar{X}$, lies $1.99s_{\bar{x}}$ *above* μ_L (66.7). This sample mean is at a distance *greater than* $1.99s_{\bar{x}}$ from any μ with a value less than μ_L. Thus, μ_L is the *lower confidence limit*, and any value of μ below this limit is considered to be "sufficiently unlikely." Similarly, $\bar{X}$ lies $1.99s_{\bar{x}}$ *below* μ_U (68.1) and would clearly be at a distance greater than $1.99s_{\bar{x}}$ from any μ greater than μ_U. Thus, μ_U is the *upper confidence limit*, and any value of μ above this limit is considered to be "sufficiently unlikely." Consequently, any null hypothesis about μ falling within the interval 66.7–68.1 should be retained; any null hypothesis specifying a value of μ that falls outside this interval should be rejected.

The confidence interval corresponding to the .01 criterion of significance is called the *99% confidence interval*. In the height experiment, the critical value of t for $df = 99$ and the .01 criterion is 2.62. Thus, the 99% confidence interval is equal to

$$67.4 - 2.62(.35) \leq \mu \leq 67.4 + 2.62(.35)$$
$$67.4 - .92 \leq \mu \leq 67.4 + .92$$
$$66.48 \leq \mu \leq 68.32$$

These limits are illustrated in the bottom part of Figure 9.7. Note that the 99% confidence interval is *wider* than the 95% confidence interval. You can be more sure that the population mean falls in the 99% confidence interval, but you must pay a price for this added

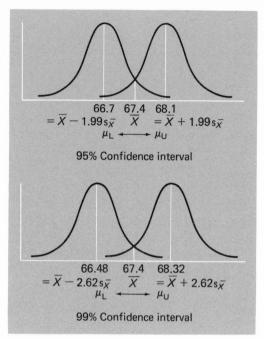

66.7   67.4   68.1
$= \bar{X} - 1.99 s_{\bar{X}}$ $\bar{X}$ $= \bar{X} + 1.99 s_{\bar{X}}$
μ_L ⟵⟶ μ_U

95% Confidence interval

66.48   67.4   68.32
$= \bar{X} - 2.62 s_{\bar{X}}$ $\bar{X}$ $= \bar{X} + 2.62 s_{\bar{X}}$
μ_L ⟵⟶ μ_U

99% Confidence interval

Figure 9.7   *95% and 99% confidence intervals for hypotheses about* μ *when* σ *is not known,* $\bar{X} = 67.4$, *s = 3.5, and N = 100.* (*Critical value of t for 99 df is 1.99 for* ɑ *= .05 and 2.62 for* ɑ *= .01.*)

confidence—the interval is larger and hence μ is less precisely estimated. (Thus, you could be absolutely certain that the population mean falls within the "confidence interval" $-\infty \leq \mu \leq +\infty$, but such a statement is useless—it tells nothing that you did not know beforehand.) For ɑ = .05, you can be 95% confident that the true population mean falls within the stated interval (narrower, hence more informative) ; for ɑ = .01, you can be 99% confident (more sure) that the true population mean falls within the stated interval (wider, hence less informative).

Be sure to note that the critical value of *t* used in the computation of confidence intervals depends on *df* (and hence *N*) for the particular problem. For example, suppose that $N = 15$, $\bar{X} = 18.80$, and $s = 3.50$. Then $s_{\bar{X}} = 3.50/\sqrt{15}$ or .90. If the 95% confidence interval is desired, the critical value of *t* for *df* = 14 is equal to 2.15, and the interval is

$$\bar{X} - t s_{\bar{X}} \leq \mu \leq \bar{X} + t s_{\bar{X}}$$
$$18.80 - 2.15(.90) \leq \mu \leq 18.80 + 2.15(.90)$$
$$18.80 - 1.94 \leq \mu \leq 18.80 + 1.94$$
$$16.86 \leq \mu \leq 20.74$$

If σ is known, use the critical z values (1.96 for the 95% confidence interval, 2.58 for the 99% confidence interval) instead of *t*.

THE STANDARD ERROR OF A PROPORTION

There is one other useful statistical procedure which involves the mean of a single population. Suppose that a few weeks before an election, a worried politician takes a poll by drawing a random sample of 400 registered voters and finds that 53% intend to vote for him, while 47% prefer his opponent.* He is pleased to observe that he has the support of more than half of the sample, but he knows that the sample may not be an accurate indicator of the population because an unrepresentative number of his supporters may have been included by chance. What should he conclude about his prospects in the election?

This problem is similar to the preceding ones in this chapter in that there is one population in which the politician is interested (registered voters) and he wishes to draw an inference about the mean of this population based on data obtained from a sample. In particular, he would like to know if the percent supporting him is greater than 50% (in which case he will win the election). There is one important difference in the present situation, however: his data are in terms of percents or *proportions*. Thus, while the general strategy is the same as in the case of the standard error of the mean, special techniques must be used that are designed to handle proportions. The needed formula is

$$z = \frac{p - \pi}{\sqrt{\pi(1 - \pi)/N}}$$

where

> p = proportion observed in the sample
> π = hypothesized value of the *population* proportion
> N = number of people in the sample

The denominator of this formula, $\sqrt{\pi(1 - \pi)/N}$, is the *standard error of a proportion*, symbolized by σ_p. It serves a similar purpose to the standard error of the mean, but it is a measure of the variability of *proportions* in samples drawn at random from a population

---

* For simplicity, we assume that there are no undecided or "won't say" voters in this sample. There will be such in reality, and their omission makes the dangerous assumption that they will break the same way as those who make a choice.

wherein the proportion in question is π. Note that since the population π is specified, and the σ of the population in question is known (it can be proved equal to $\sqrt{\pi(1-\pi)}$), the results are referred to the z table. In the case of the anxious politician, the critical population value in which he is interested is .50 ; he wants to know whether the observed sample proportion of .53 is sufficiently different from .50 to enable him to conclude that a majority of the population of voters will vote for him. Thus,

$\pi =$ hypothesized population proportion $= .50$
$p =$ proportion observed in sample $= .53$

The statistical analysis is as follows :

H_0: $\pi = .50$
H_1: $\pi \neq .50$
    $a = .05$

$$z = \frac{.53 - .50}{\sqrt{.50(1-.50)/400}} = \frac{.03}{.025} = 1.20$$

The value of 1.20 is smaller than the critical z value of 1.96 needed to reject H_0. Therefore, the politician *cannot* conclude that he will win the election; it is "not sufficiently unlikely" for 53% of a sample of 400 voters to support him even if only 50% of the population will vote for him. The politician will therefore have a nervous few weeks until the votes are in, and he may well conclude from these results that he should increase his campaigning efforts. The results are illustrated in Figure 9.8.

Figure 9.8 *Sampling distribution of p when π = .50 and N = 400.*

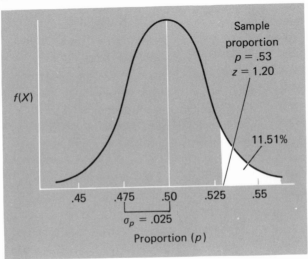

There is one important word of caution regarding the use of the standard error of the proportion. When the hypothesized value of the population proportion (π) is quite large or small and the sample size is small, the sampling distribution of proportions is *not* normal. Hence, computing a z value and referring it to the normal curve table will yield incorrect results in these situations. As a rough guide, you should *not* use this procedure when either $N\pi$ or $N(1-\pi)$ is less than 5. For example, if you wish to test the null hypothesis that $\pi = .85$ with a sample size of 20,

$$N\pi = (.85)(20) = 17 \quad \text{and} \quad N(1-\pi) = (.15)(20) = 3.$$

Since $N(1-\pi)$ is less than 5, the procedures discussed in this section should *not* be used; the tested value of π is too extreme and the sample size is too small to give an accurate answer. You should either use the appropriate statistical techniques to deal with this situation (the binomial distribution, discussed in more advanced texts) or increase the size of your sample.

CONFIDENCE INTERVALS FOR π

Suppose you wish to determine all reasonably likely values of the population proportion (π) using the .05 criterion of significance. Since the sampling distribution is normal, the confidence interval is

$$(p - 1.96\sigma_p) \leq \pi \leq (p + 1.96\sigma_p)$$

There is one difficulty: σ_p depends on the value of π, which is not known. In the previous section, a value of π was specified by H_0 and used in the calculation of σ_p, but there is no single hypothesized value of π in interval estimation.

One possibility, which yields a fairly good approximation *if the sample size is large*, is to use p as an estimate of π. With a large sample, p is unlikely to be so far from π as to greatly affect σ_p and hence the size of the confidence interval. In the case of the nervous politician in the previous section, $p = .53$; and $\sqrt{.53(1-.53)/N}$ is equal to .025. Then, the 95% confidence interval is equal to

$$.53 - 1.96(.025) \leq \pi \leq .53 + 1.96(.025)$$
$$.53 - .049 \leq \pi \leq .53 + .049$$
$$.48 \leq \pi \leq .58$$

Since the sample size is large ($N = 400$), the politician can state with 95% confidence that between approximately 48% and 58% of the population will vote for him. Note that the value of .50 *is* within the confidence interval, as would be expected from the fact that the null hypothesis that $\pi = .50$ was retained. To determine the 99% confidence interval, the procedure would be the same except that the z value of 2.58 would be used instead of 1.96.

"Exact" solutions for the confidence interval for π (which do not require using p as an estimate of π and hence large samples) do exist, but involve techniques that are beyond the scope of this book.

ONE-TAILED TESTS OF SIGNIFICANCE

The statistical tests discussed earlier in this chapter (see for example Figures 9.3, 9.4, and 9.5) are called *two-tailed* tests of significance. This means that H_0 is rejected if a z or t value is obtained that is either extremely high (far up in the upper tail of the curve) *or* extremely low (far down in the lower tail of the curve). Consequently, there is a rejection area equal to $a/2$ in each tail of the distribution. As we have seen, the corresponding null and alternative hypotheses are:

H_0: $\mu =$ a specific value
H_1: $\mu \neq$ this value (that is, μ is greater than the value specified by H_0 *or* is less than this value)

Let us suppose that a researcher argues as follows: "My theory predicts that the mean of population X is *less than* 100." (For example, he may predict that children from homes in which there are few books and little emphasis on verbal skills are below average on a standardized test of intelligence.) "It's all the same to me whether μ_x is equal to 100 or much greater than 100; my theory is disconfirmed in either case. If I use a two-tailed test, I am forced to devote $2\frac{1}{2}$% of my rejection region (using $a = .05$) to an outcome which is meaningless insofar as my theory is concerned. Instead, I will place the entire 5% rejection region in the lower tail of the curve." (See Figure 9.9.) "In other words, I will test the following null and alternative hypotheses:

H_0: $\mu_x \geq 100$
H_1: $\mu_x < 100$
$a = .05$

"Note that I am being properly conservative by assuming at the outset that my results are due to chance (the null hypothesis); only if I obtain a result very unlikely to be true if H_0 is true will I conclude that my theory is supported (switch to the alternative

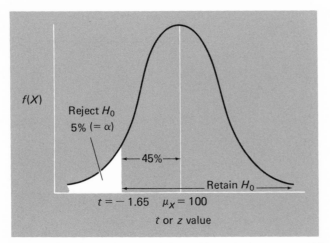

Figure 9.9 *One-tailed test of the null hypothesis that $\mu_X \geq 100$ against the alternative that $\mu_X < 100$.*

hypothesis). Looking at the t table for $df = 500$ (since I have $N = 501$ and σ is not known), I see that a t value of -1.65 places 5% of the distribution in the lower tail. Therefore, I will retain H_0 if t is anywhere between -1.65 and $+\infty$ and reject H_0 if t is -1.65 or less. Since a two-tailed test of significance would lead to rejection of H_0 only if t were -1.96 or less (or $+1.96$ or more), the one-tailed value of -1.65 quite properly makes it easier for me to reject H_0 (and conclude that my theory is supported) when the results are in the direction that I have predicted."

Similar reasoning could be applied if the researcher claimed to be interested *only* in the outcome that the population mean was *greater* than 100. The entire rejection region would be placed at the upper end of the curve, and H_0 would be retained if t were between $-\infty$ and $+1.65$ and rejected if t were $+1.65$ or more.

Some statisticians recommend that one-tailed tests be used in research in the behavioral sciences, not only for the techniques described in this chapter but also for procedures discussed in subsequent chapters. You are therefore likely to encounter them when you consult the professional literature. It is our belief, however, that one-tailed tests should *not* be used in place of two-tailed tests of significance. One reason for this position is that the user of one-tailed tests is placed in an embarrassing position if extreme results are obtained in the direction opposite to the one expected. For example, suppose the researcher in the example at the beginning of this section finds that t is equal to *plus* 4.0 (that is, the mean of the sample is much *larger* than 100). *Both* the null and alternative hypotheses are different in the case of the one-tailed test, and the researcher must retain the new null hypothesis (that $\mu_X \geq 100$). That is, he must do the same thing when t is equal to $+4.0$ as when t is equal to 0.0 or -1.60, namely conclude that 100 is (among others, including for example 150) a reasonably likely value of the population mean. Since the probability of obtaining a t value of $+4.0$ or more, if the population mean is in fact 100, is approximately .00003, the conclusion forced upon the researcher by the one-tailed test is exceedingly awkward. Yet there is little else he can do; he has opted to place the entire rejection region at the lower tail of the distribution, and he must stick with (perhaps we should say "be stuck with") that decision lest he violate the .05 criterion of significance. For if he changes his mind after the results are in and decides to reject H_0 when t is equal to $+1.96$ or more, he is in fact using the .075 criterion. That is, he rejects H_0 when t is equal to or less than the -1.65 specified by the one-tailed test (which will happen by chance .05 of the time when H_0 is true), and when t is $+1.96$ or more (which will happen by chance .025 of the time when H_0 is true). Thus, his probability of making a Type I error is *not* .05; it is .075—he is more likely to reject a true H_0 than is the case when the .05 criterion is used.

Regardless of how the researcher ultimately resolves this unpleasant dilemma, he will undoubtedly ask himself why children in "less verbal" environments are superior in intelligence—thereby making clear that he really was operating on a two-tailed basis all along, since extreme results at the upper tail of the curve *are* in fact different to him than results near the hypothesized value of the population mean. In fact, in almost all situations in the behavioral sciences, extreme results in either direction are of interest. Even if results in one direction are inconsistent with a given theory, they may suggest new lines of thought. Also, theories change over time, but published results remain immutable on the printed page. Those perusing the professional journals (perhaps years after the article was written) should not be bound by what theory the original researcher happened to believe, as happens with the one-tailed test which arbitrarily equates extreme results in one direction with nonextreme results close or equal to the value specified by the null hypothesis.

We recognize, however, that our position concerning one-tailed tests will not meet with universal support. To the researcher who wishes to use a one-tailed test, therefore, we recommend that if the result would have been significant in the "other" direction, the *experiment should be repeated before conclusions are drawn*.

The extra effort involved is the consequence of obtaining extreme results in the direction opposite to that predicted when a one-tailed test is used.

To sum up: The user of one-tailed tests (for example, that H_1 is $\mu < 100$) must renounce all interest in results of any magnitude in the opposite direction (for example, that $\mu > 100$). Since extreme results in the "opposite" direction invariably have important implications for the state of scientific knowledge, implications which are quite different than had nonextreme results been obtained, the use of one-tailed tests is not recommended. We do not question the *statistical* validity of such tests, but rather their operational validity in the present state of behavioral science. If a one-tailed test is used and results are obtained in the "opposite" direction which would have been significant, the experiment should be repeated before conclusions are drawn.

SUMMARY

The first step in *hypothesis testing* is to set up the *null hypothesis* (H_0) and the *alternative hypothesis* (H_1). Without measuring the entire population, it is impossible to be certain as to which of these two hypotheses is correct. A *criterion of significance* is then selected; next, the appropriate statistical procedure is used to compute the probability of obtaining the observed results *if H_0 is true*. Then, H_0 is rejected if the probability is "sufficiently small" in terms of the chosen significance criterion (that is, if the computed statistic *exceeds or equals* the "critical value" obtained from the appropriate table); otherwise, H_0 is retained.

The decision based on the statistical analysis runs the risk of one of two kinds of error:

Type I error: Rejecting H_0 when H_0 is actually true. The probability of committing this error is equal to the chosen criterion of significance.

Type II error: Failing to reject (retaining) H_0 when H_0 is actually false. The probability of committing this error is not so conveniently determined (see Chapter 13).

To test null hypotheses about the *mean of one population*, first compute the estimated *standard error of the mean*, a measure of how accurate the sample mean is likely to be as an estimate of the population mean. Then, use the standard error of the mean to determine whether the difference between the observed sample mean and the hypothesized value of the population mean is or is not "sufficiently unlikely" to occur if H_0 is true (that is, if the hypothesized value of the population mean is correct). If the population standard deviation is known, use the normal curve model; if it is *not* known, use the *t distributions*.

To test null hypotheses about the *proportion of one population*, the procedure is similar except that the *standard error of the proportion* is used and the results are referred to the normal curve model.

Confidence intervals may be used to determine all reasonably likely values of a population parameter, such as a mean or proportion.

10 testing hypotheses about the difference between the means of two populations

In the preceding chapter we looked at techniques for drawing inferences about the mean of one population. More often, however, scientists wish to draw inferences about differences between two or more populations. For example, a social psychologist may want to know if the mean of the population of white Americans on a pencil-and-paper test of musical aptitude is equal to the mean of the population of black Americans. Or, an experimental psychologist may wish to find out whether rats perform better on a discrimination learning task to gain a reward or to avoid punishment—that is, whether the mean number of correct responses is greater for the "reward" population or for the "punishment" population. In both cases, the question of interest concerns a comparison between the means of *two* samples.

In this chapter our objective will be to develop techniques needed to draw inferences about the *difference between the means of two populations* ($\mu_1 - \mu_2$), using a general strategy similar to that of the preceding chapter. Procedures for drawing inferences about the means of more than two populations, or about other aspects of two or more populations, will be considered in later chapters.

THE STANDARD ERROR OF THE DIFFERENCE

Suppose that you wish to conduct a research study to determine whether or not the use of caffeine improves performance on a college mathematics examination. To test this hypothesis, you obtain two *random* samples of college students taking mathematics, an *experimental group* and a *control group*. (Random sampling is one good way of avoiding groups greatly different in ability, motivation, or any other variable which might well obscure the

131

effects of the caffeine.) You then give all members of the experimental group a small dose of caffeine one hour prior to the test. At the same time, the control group is given a placebo (a pill which, unknown to them, has no biochemical effect whatsoever) to control for the possibility that administering *any* pill will affect the students' performance (for example, they may become more "psychologically" alert and obtain higher scores). Thus, the control group serves as a baseline against which the performance of the experimental group can be evaluated. Finally, the mean test scores of the two groups are compared. Suppose that the mean of the experimental group equals 81 and the mean of the control group equals 78. Should you conclude that caffeine is effective in improving the test scores?

As was shown in Chapter 9, any sample mean is almost never exactly equal to the population mean because of sampling error—those cases which happened, by chance, to be included in the sample. Therefore, the two populations in question (a hypothetically infinite number of test-takers given caffeine and a hypothetically infinite number of test-takers given a placebo) may well have equal means even though the sample means are different. Consequently, you cannot tell what conclusion to reach just by looking at the sample means. You need additional information: namely, whether the three-point difference between 81 and 78 is likely to be a trustworthy indication that the population means are different (and that therefore caffeine *has* an effect), or whether this difference may be due solely to which cases happened to fall in each sample, in which case you should *not* conclude that caffeine has an effect.

How can this additional information be obtained? In Chapter 9 we saw that the variability of the sampling distribution of means drawn from one population, $s_{\bar{X}}$, provides information as to how trustworthy is any one $\bar{X}$ as an estimate of μ. The present problem (how trustworthy is any one *difference*, $\bar{X}_1 - \bar{X}_2$, as an estimate of the difference between the population means, $\mu_1 - \mu_2$) is also solved by making use of an appropriate sampling distribution—the sampling distribution of the *difference* between *two* sample means.

Let us suppose that a variable is normally distributed in each of two defined populations. Let us further suppose that the populations have equal means and equal variances (that is, $\mu_1 = \mu_2$ and $\sigma_1^2 = \sigma_2^2$). If many pairs of random samples of equal size were actually drawn from the two populations, a distribution of differences between the paired means could be established empirically. For each pair, you would subtract the mean of the second sample

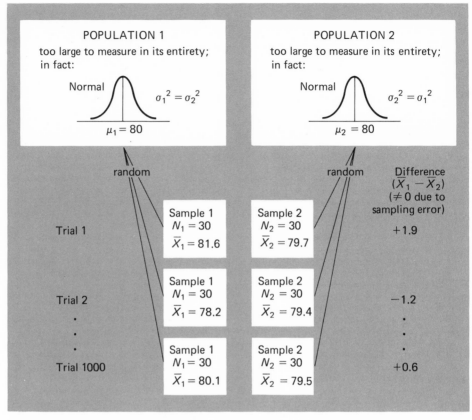

Figure 10.1 *Illustration of procedure for obtaining the empirical sampling distribution of differences between two means (N = 30).*

$(\bar{X}_2)$ from the mean of the first sample $(\bar{X}_1)$. Since the means of the two populations are equal, any difference between the sample means must be due solely to sampling error.

Figure 10.1 illustrates the procedure needed to obtain an empirical sampling distribution of 1000 differences in the case where $\mu_1 = \mu_2 = 80$ and the size of each random sample is 30. The resulting frequency distribution is shown in Table 10.1, and the frequency polygon plotted from this distribution is shown in Figure 10.2. As expected, the sample means are not in general exactly equal to 80, and the differences between pairs of means are not in general exactly equal to zero, due to sampling fluctuations. Notice that the distribution is approximately symmetric, which indicates that chance differences are equally likely to occur in either direction—that is, $(\bar{X}_1 - \bar{X}_2) > (\mu_1 - \mu_2)$ or $(\bar{X}_1 - \bar{X}_2) < (\mu_1 - \mu_2)$. Thus, the mean of the distribution of differences tends to be equal to $\mu_1 - \mu_2$, which in this case is zero.

Table 10.1 *Empirical sampling distribution of 1000 differences between pairs of sample means (N = 30) drawn from two populations where* $\mu_1 = \mu_2 = 80$ *(hypothetical data)*

difference between sample means $(\bar{X}_1 - \bar{X}_2)$	number of samples (f)
Greater than +11.49	0
+10.50 to +11.49	1
+9.50 to +10.49	0
+8.50 to +9.49	1
+7.50 to +8.49	4
+6.50 to +7.49	7
+5.50 to +6.49	21
+4.50 to +5.49	32
+3.50 to +4.49	54
+2.50 to +3.49	77
+1.50 to +2.49	107
+.50 to +1.49	122
−.50 to +.49	153
−1.50 to −.51	114
−2.50 to −1.51	95
−3.50 to −2.51	82
−4.50 to −3.51	60
−5.50 to −4.51	31
−6.50 to −5.51	22
−7.50 to −6.51	10
−8.50 to −7.51	4
−9.50 to −8.51	1
−10.50 to −9.51	2
Smaller than −10.50	0
Total	1000

Given the data in Table 10.1, you could actually calculate the standard deviation of the distribution of differences for the given sample size by using the usual formula for the standard deviation (Chapter 5). This standard deviation would tell you how much, on the average, a given $\bar{X}_1 - \bar{X}_2$ is likely to differ from the "true" difference of zero (the central point of the distribution of differences). Consequently, it would indicate how trustworthy is the single difference that you have on hand, and it is therefore called the *standard error of the difference*. A relatively *large* standard error of the difference would indicate that any single difference

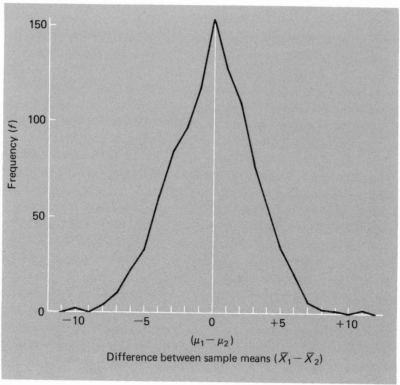

Figure 10.2 *Frequency polygon of data in Table 10.1.*

between a pair of sample means must be viewed with grave suspicion; since large discrepancies between $\bar{X}_1$ and $\bar{X}_2$ are likely *even if they come from populations with equal means*, you need a sizable difference indeed before you can safely conclude that μ_1 is not equal to μ_2. If, on the other hand, the standard error of the difference is relatively *small*, you can place more confidence in any one sample difference as an estimate of the population difference. Large discrepancies between $\bar{X}_1$ and $\bar{X}_2$ are *not* likely to occur (and mislead you) if in fact $\mu_1 = \mu_2$, so a large difference between the sample means *is* a trustworthy sign that the population means differ.

In practice, the behavioral scientist never has many paired random samples of a given size; he has just one. Therefore, it is impossible to compute the standard error of the difference empirically, and (just as in Chapter 9) you must make use of estimation procedures developed by statisticians. In the present situation you will need techniques that enable you to *estimate* the standard error of the difference based on the *one* pair of samples at your disposal.

ESTIMATING THE STANDARD ERROR OF THE DIFFERENCE

As was the case with $s_{\bar{x}}$, developing the estimation formula for the standard error of the difference by means of a formal proof is beyond the scope of this book. Instead, we will attempt to explain the formula in nonmathematical terms.

Since it is assumed that the two population variances are equal, the variances of each of the two samples may be combined into a single estimate of the common value of σ^2. If the sizes of the two samples are exactly equal, you can obtain this combined or *pooled* estimate by simply computing the average of the two sample variances. Often, however, the sample sizes are not equal, in which case greater weight must be given to the larger sized sample; a *weighted* average must be computed. The general formula for the *pooled variance* (symbolized by s^2_{pooled}), which may be used for equal or unequal sample sizes, is:

$$s^2_{pooled} = \frac{(N_1 - 1)s_1^2 + (N_2 - 1)s_2^2}{N_1 + N_2 - 2}$$

where s_1^2 = variance of sample 1
s_2^2 = variance of sample 2
N_1 = number of cases in sample 1
N_2 = number of cases in sample 2

Then, the estimation formula for the standard error of the difference is:

$$s_{\bar{x}_1 - \bar{x}_2} = \sqrt{\frac{s^2_{pooled}}{N_1} + \frac{s^2_{pooled}}{N_2}} = \sqrt{s^2_{pooled}\left(\frac{1}{N_1} + \frac{1}{N_2}\right)}$$

where $s_{\bar{x}_1 - \bar{x}_2}$ = estimated standard error of the difference

This formula should not look totally unfamiliar. We saw in Chapter 9 that the estimation formula for the standard error of the mean ($s_{\bar{x}}$) is equal to $s/\sqrt{N}$. Consequently, $s_{\bar{x}}^2$ is equal to s^2/N. Thus, the estimation formula for the standard error of the difference between two means is closely related to the estimated standard error of each of the individual sample means. As you can see from the formula, with the difference subject to sampling error from *two* means, the sampling error variance reflects both sources of error; in addition, it uses a more stable estimate of the population variance by pooling the variance information from the two samples.

Notice also that the standard error of the difference gets smaller as the size of each sample increases. Differences based on large samples will have smaller sampling error (and are thus likely to be more accurate estimates of the population difference) than are differences based on small samples, just as is the case for a single mean.

The two steps given above can be combined into a single formula, as follows*:

$$s_{\bar{X}_1 - \bar{X}_2} = \sqrt{\frac{(N_1 - 1)s_1^2 + (N_2 - 1)s_2^2}{N_1 + N_2 - 2} \left(\frac{1}{N_1} + \frac{1}{N_2}\right)}$$

THE _t_ TEST FOR TWO SAMPLE MEANS

Once the estimated standard error of the difference has been computed, a _t_ value (and therefore a probability value) may be assigned to any single obtained difference, as follows:

$$t = \frac{(\bar{X}_1 - \bar{X}_2) - (\mu_1 - \mu_2)}{s_{\bar{X}_1 - \bar{X}_2}}$$

Notice that the two sample means yield a single difference score, $\bar{X}_1 - \bar{X}_2$, which is compared to the null-hypothesized mean of the difference scores, $\mu_1 - \mu_2$. It is possible to test any hypothesized difference between the mean of two populations; for example, you could test the hypothesis that the mean of the first population is 20 points greater than the mean of the second population by setting $\mu_1 - \mu_2$ equal to 20 in this equation. Much more often than not, however, you will want to test the hypothesis that the means of the two populations are equal (that is, $\mu_1 = \mu_2$ or $\mu_1 - \mu_2 = 0$). In this situation, the _t_ formula can be simplified to

$$t = \frac{\bar{X}_1 - \bar{X}_2}{s_{\bar{X}_1 - \bar{X}_2}}$$

---

* Since

$$s^2 = \frac{\sum (X - \bar{X})^2}{N - 1}$$

and therefore $(N - 1)s^2 = \sum (X - \bar{X})^2$, this formula may also be written as

$$s_{\bar{X}_1 - \bar{X}_2} = \sqrt{\frac{\sum (X - \bar{X}_1)^2 + \sum (X - \bar{X}_2)^2}{N_1 + N_2 - 2} \left(\frac{1}{N_1} + \frac{1}{N_2}\right)}$$

$$= \sqrt{\frac{\sum X_1^2 + \sum X_2^2 - \dfrac{(\sum X_1)^2}{N_1} - \dfrac{(\sum X_2)^2}{N_2}}{N_1 + N_2 - 2} \left(\frac{1}{N_1} + \frac{1}{N_2}\right)}$$

or

$$t = \frac{\bar{X}_1 - \bar{X}_2}{\sqrt{\dfrac{(N_1 - 1)s_1^2 + (N_2 - 1)s_2^2}{N_1 + N_2 - 2}\left(\dfrac{1}{N_1} + \dfrac{1}{N_2}\right)}} \qquad df = N_1 + N_2 - 2$$

The degrees of freedom upon which this t value is based (needed to obtain the critical value of t from the t table) is determined from the fact that the population σ^2 is estimated using $N_1 - 1$ degrees of freedom from sample 1 and $N_2 - 1$ degrees of freedom from sample 2 (see Chapter 9). Therefore, the estimate is based on the combined degrees of freedom:

$$(N_1 - 1) + (N_2 - 1) \quad \text{or} \quad N_1 + N_2 - 2$$

Returning to the problem of caffeine and mathematics test scores posed at the beginning of this chapter, the first step in the statistical analysis consists of stating the hypotheses and establishing a significance criterion or a:

H_0: $\mu_{exp} = \mu_{con}$
H_1: $\mu_{exp} \neq \mu_{con}$
$\qquad a = .05$

H_0 states that the experimental and control samples come from populations with equal means, while H_1 states that these two samples come from populations with different means. The customary .05 criterion of significance is specified.

Next, let us suppose that the experimental results are as follows:

experimental group   control group

$N_1 = 22$ $\qquad\qquad$ $N_2 = 20$
$\bar{X}_1 = 81.0$ $\qquad\quad$ $\bar{X}_2 = 78.0$
$s_1^2 = 12.0$ $\qquad\quad$ $s_2^2 = 10.0$

The estimated standard error of the difference is equal to:

$$s_{\bar{X}_1 - \bar{X}_2} = \sqrt{\frac{(21)(12) + (19)(10)}{22 + 20 - 2}\left(\frac{1}{22} + \frac{1}{20}\right)} = \sqrt{(11.05)\left(\frac{1}{22} + \frac{1}{20}\right)}$$

$$= 1.03$$

and the t value is equal to:

$$t = \frac{81.0 - 78.0}{1.03} = 2.91, \qquad df = 22 + 20 - 2 = 40$$

Referring to the t table for 40 degrees of freedom, a t value greater in absolute value than 2.02 is needed to justify rejection of H_0 using the .05 criterion of significance. Since the obtained

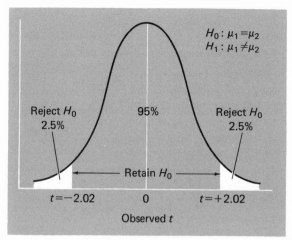

Figure 10.3 *Acceptance and rejection regions for caffeine experiment in the t distribution for df = 40.*

value of t is in fact greater in absolute value than the critical value obtained from the table, it is *not* likely that the obtained difference between the sample means is due to sampling error. Therefore, you reject H_0 and conclude that the experimental and control groups are random samples from populations with *different* means, and more specifically that caffeine *has* a positive effect on mathematics test scores.

The rejection areas for this experiment are illustrated in Figure 10.3.

MEASURES OF THE STRENGTH OF THE RELATIONSHIP BETWEEN THE TWO VARIABLES

It is very desirable to convert statistically significant values of t for the difference between two means to an index which will show the *strength* of the relationship between the two variables. For example, the statistically significant value of t obtained in the caffeine experiment in the preceding section indicates that caffeine has a positive effect on mathematics test scores, but it does *not* indicate whether the relationship between the presence or absence of caffeine and test scores is weak, moderate, or strong. To obtain this important information about the strength of the relationship between caffeine and test scores, it is necessary to convert the obtained value of t to a statistic called the point biserial correlation coefficient. Since we have not yet considered the fundamentals of correlation, however, we will defer further discussion of this important topic until Chapter 12.

CONFIDENCE INTERVALS FOR $\mu_1 - \mu_2$

The use of confidence intervals to determine all reasonably likely values of the mean of a *single* population was discussed in Chapter 9. In a similar fashion, all reasonably likely values of the *difference between two population means* can be found by using the following confidence interval:

$$[(\bar{X}_1 - \bar{X}_2) - ts_{\bar{X}_1 - \bar{X}_2}] \leq \mu_1 - \mu_2 \leq [(\bar{X}_1 - \bar{X}_2) + ts_{\bar{X}_1 - \bar{X}_2}]$$

where $t =$ critical value obtained from the t table for the specified criterion of significance and df

Note that since this confidence interval pertains to the difference between two population means, the appropriate error term is the standard error of the difference $(s_{\bar{X}_1 - \bar{X}_2})$.

For example, in the caffeine experiment, the observed sample means were 81.0 for the experimental group and 78.0 for the control group; the standard error of the difference was 1.03; and the critical value of t for $df = 40$ and $a = .05$ was 2.02. The 95% confidence interval for the difference between the population means is:

$$[(81.0 - 78.0) - (2.02)(1.03)] \leq \mu_1 - \mu_2$$
$$\leq [(81.0 - 78.0) + (2.02)(1.03)]$$
$$(3.0 - 2.08) \leq \mu_1 - \mu_2 \leq (3.0 + 2.08)$$
$$.92 \leq \mu_1 - \mu_2 \leq 5.08$$

Thus, you can state with 95% confidence that the mean of the experimental population on the mathematics test is included in the interval that runs from .92 points to 5.08 points greater than the mean of the control population. Note that zero is *not* in this interval, indicating that it is *not* likely that $\mu_1 = \mu_2$.

To determine the 99% confidence interval, the procedure would be the same except that the critical value of t for 40 df for $a = .01$, which is 2.70, would be used instead of the t value for 40 df for $a = .05$ of 2.02.

USING THE t TEST FOR TWO SAMPLE MEANS: SOME GENERAL CONSIDERATIONS

IMPLICATIONS OF RETAINING H_0

Suppose that a different caffeine experiment yielded an obtained t value of 1.03 for $df = 40$. Since 1.03 is smaller than the critical value of t obtained from the table, you would decide to retain H_0.

This does *not* imply, however, that you have shown that the population means are equal. You do not know how likely your decision to retain H_0 is to be wrong (that is, how likely you are to make a Type II error), so you cannot make a definite statement as to your confidence that the population means are equal. Unless you make use of techniques for determining the probability of a Type II error (see Chapter 13), therefore, you must limit yourself to a cautious statement such as "there is not sufficient reason to reject the hypothesis that caffeine has no effect."

IMPLICATIONS OF REJECTING H_0

When the results of the statistical analysis indicate that you should *reject H_0* (that is, the results are *statistically significant*), you may then conclude that caffeine has an effect. The probability of rejecting H_0 erroneously (committing a Type I error) *is* known; it is equal to a (or .05 in the preceding caffeine experiment). Thus, you may be 95% confident that your decision to reject H_0 is correct.

Notice, however, that there is still a 5% chance that you have made a Type I error and that the population means actually are equal. Therefore, it is usually *not* a good idea to regard any one such statistical finding as completely conclusive, and psychologists would properly wait to see if repetitions (*replications*) of the caffeine experiment also indicated that caffeine did have an effect before reaching firm conclusions.

VIOLATIONS OF THE UNDERLYING ASSUMPTIONS

In theory, the use of the t test discussed in this chapter is justified only if two assumptions are met:

1.   The variable is normally distributed within each population.
2.   The variances of the two populations are equal ($\sigma_1^2 = \sigma_2^2$).

In practice, however, you should not be deterred from using the two-tailed t test even if the assumption concerning normality is not exactly met. This test is *robust* with regard to this assumption —it gives fairly accurate results even if the assumption is not satisfied.

The assumption concerning the equality of the two population variances can be practically ignored as well *if* the two sample sizes are equal. If, however, the sample sizes are fairly unequal (say, the larger is more than 1.5 times greater than the smaller) *and* the population variances are markedly unequal, the t test may well yield erroneous results. Therefore, if there is reason to believe that the population variances differ substantially, it is a good idea to obtain approximately equal sample sizes so that any differences between the population variances will not bias the results obtained from the t test. Otherwise, there are available special tests to use when you have unequal variances with unequal sample sizes.

THE t TEST FOR MATCHED SAMPLES

Suppose that you want to test the hypothesis that in families with two children, the first-born is more introverted than the second-born. Once again, the question of interest concerns a comparison between the means of two populations, and the null and alternative hypotheses are as follows:

H_0: $\mu_{\text{first-born}} = \mu_{\text{second-born}}$
H_1: $\mu_{\text{first-born}} \neq \mu_{\text{second-born}}$

To obtain your samples, you randomly select *matched pairs* of children, with each pair consisting of the first-born child and the second-born child from a given family. You then administer an appropriate measure of introversion. The general format of the resulting data would be as follows:

pair	first-born (X_1)		second-born (X_2)
1	65	←—matched—→	61
2	48	←—matched—→	42
3	63	←—matched—→	66
⋮	⋮		⋮
N	66	←—matched—→	69

Thus, the first-born child from family 1 has an introversion score of 65, and the second-born child from family 1 has an introversion score of 61. Similarly, each pair of children is matched by virtue of coming from the same family. This procedure is different from the caffeine experiment discussed previously, which involved *independent* random samples (that is, there was no connection between any specific individual in sample 1 and any specific individual in sample 2). In fact, the statistical analysis described previously in this chapter is *not* suitable for matched samples.

The procedures needed to analyze matched pair data are similar to those already developed in Chapter 9. Since each subject in sample 1 has a paired counterpart in sample 2, you can subtract each X_2 from each X_1 and obtain a difference score (denoted by D). Then, you can use the techniques of the previous chapter for drawing inferences about the mean of one population—the mean of the population of difference scores.

To illustrate this procedure, let us suppose that you obtain 10 matched pairs of children, and that the results are as shown in Table 10.2. The difference scores are shown in the last column

Table 10.2 *Introversion scores for 10 matched pairs of siblings (hypothetical data)*

pair	first-born (X_1)	second-born (X_2)	$D = (X_1 - X_2)$
1	65	61	+4
2	48	42	+6
3	63	66	−3
4	52	52	0
5	61	47	+14
6	53	58	−5
7	63	65	−2
8	70	62	+8
9	65	64	+1
10	66	69	−3
$\sum$	606	586	+20
Mean	60.6	58.6	2.0

of the table; note that the *sign* of each difference score must be retained. The null and alternative hypotheses are now as follows:

$$H_0: \quad \mu_D = 0$$
$$H_1: \quad \mu_D \neq 0$$

H_0 states that the mean of the population of difference scores is zero, and H_1 states that the mean of the population of difference scores is not zero. The next step is to compute the mean difference score obtained from the sample, $\bar{D}$:

$$\bar{D} = \frac{\sum D}{N} = \frac{20}{10} = 2.0$$

Note that N is not the number of subjects, but the number of *pairs* of subjects. As is shown in Table 10.2, $\bar{D}$ is equal to $\bar{X}_1 - \bar{X}_2$. Thus, testing the hypothesis that $\mu_D = 0$ is equivalent to testing the hypothesis that $\mu_1 - \mu_2 = 0$, or that $\mu_1 = \mu_2$.

To test the hypothesis about the mean of one population, you compare the sample mean to the hypothesized value of the population mean and divide by the standard error of the mean, as follows:

$$t = \frac{\bar{D} - \mu_D}{\sqrt{\dfrac{s_D^2}{N}}}$$

where s_D^2 = variance of the difference scores, computed by the usual formula:

$$s_D^2 = \frac{\sum (D - \bar{D})^2}{N - 1} \quad \text{or} \quad s_D^2 = \frac{\sum D^2 - \dfrac{(\sum D)^2}{N}}{N - 1}$$

Since the hypothesized value of μ_D is zero, this formula may be simplified to:

$$t = \frac{\bar{D}}{\sqrt{s_D^2 / N}}$$

These two steps may be summarized by the following (algebraically identical) computing formula:

$$t = \frac{\sum D}{\sqrt{\dfrac{N \sum D^2 - (\sum D)^2}{N - 1}}}$$

As is normally the case in tests concerning one population mean, the number of degrees of freedom on which t is based is equal to one less than the number of scores, or

(number of *pairs*) $- 1$

If the computed t is smaller in absolute value than the value of t obtained from the table for the appropriate degrees of freedom, retain H_0; otherwise reject H_0 in favor of H_1.

Returning to the introversion study, the analysis of the data is as follows:

$$a = .05$$

$$\sum D = 20, \qquad \bar{D} = 2.0, \qquad \sum D^2 = 360$$

$$s_D^2 = \frac{360 - \dfrac{(20)^2}{10}}{9} = 35.56$$

$$t = \frac{2.0}{\sqrt{35.56/10}} = 1.06$$

Or, using the computing formula,

$$t = \frac{20}{\sqrt{\dfrac{(10)(360) - (20)^2}{9}}} = 1.06$$

The critical value of t obtained from the table for $(10-1)$ or 9 degrees of freedom is 2.26. Since the obtained t value of 1.06 is smaller in absolute value than the critical value from the table, you retain H_0 and conclude that there is *not* sufficient reason to believe that first-born and second-born children differ in introversion.

SUMMARY

To test null hypotheses about *differences between the means of two populations* using *independent* samples, first compute the estimated *standard error of the difference*, a measure of how accurate the observed difference between the two sample means is likely to be as an estimate of the difference between the population means. Then, use the standard error of the difference to determine whether the difference between the observed sample means is or is not "sufficiently unlikely" if H_0 is true, using the t distributions. With *matched* samples, compute a difference score for each pair and use the procedures for estimating the mean of the single population of difference scores; reject H_0 if this mean is "sufficiently unlikely" to be zero.

11 linear correlation and prediction

Control over the environment is achieved by man when he comes to understand, through experience, the relationship among many of the events in his life—that is, the relative frequency with which two or more phenomena occur jointly. Without such knowledge it would be impossible to make accurate predictions about future events. For example, you come to realize early in your student career that the amount of time spent in studying is related to grades. True, some students may study many hours and obtain poor grades, while others may achieve high grades despite short study periods. The general trend holds, however; in the majority of cases, you can accurately predict that little or no studying will lead to poor grades while more studying will result in better grades. In statistical terminology, the two variables of hours studying and grades are said to be co-related or *correlated*.

Many pairs of variables are correlated, while many are unrelated or *uncorrelated*. For example, sociologists have found that the income of a family is positively related to the IQ of a child in that family; the more income, the higher the child's IQ. Thus, these variables are said to be *positively correlated*. This relationship is depicted graphically in Figure 11.1, which is called a *scatter plot* (or scatter diagram) because the points scatter across the range of scores. Note that each point on the graph represents two values for one family, income (X variable) and IQ of the child (Y variable). Also, in the case of a positive correlation, the straight line summarizing the points slopes *up* from left to right. The golf enthusiast will readily accede to the fact that the number of years of play is negatively related to his golf score; the more years of practice, the fewer shots needed to complete a round of 18 holes. These two variables are said to be *negatively correlated*. This relationship is illustrated in Figure 11.2; notice that in the case of a negative correlation, the straight line summarizing the points slopes *down* from left to right. Length of big toes among male adults is un-correlated with IQ scores (see Figure 11.3).

When you can demonstrate that two variables are correlated, you can use the score of an individual on one variable to *predict* or *estimate* his score on the other variable. For example, in Figure 11.2 a fairly accurate prediction of an individual's golf score can be made from the number of years he has played golf. The more closely the two variables are related, the better the prediction is likely to be; while if two variables are uncorrelated (as in the case of toe length and IQ), you cannot accurately predict an individual's score on one of them from his score on the other. Thus, the concepts of correlation and prediction are closely related.

This chapter deals with three main topics: (1) the *description* of the relationship between two variables for a set of observations; (2) making *inferences* about the strength of the relationship between two variables in a population given the data of a sample; and (3) the *prediction* or estimation of values on one variable from observations on another variable with which it is paired. In particular, we will deal with one very common kind of relationship between two variables, namely a *linear* or straight-line relationship. That is, if values of one variable are plotted against values of the other variable on a graph, the trend of the plotted points can best be represented by a straight line. Notice that the data plotted in Figures 11.1 and 11.2 tend to fall near or on the straight line drawn through the scatter plot; this indicates that the two variables in question are highly linearly correlated. The points in Figure 11.3, on the other hand, are scattered randomly throughout the graph and cannot be represented well by a straight line; these two variables are linearly uncorrelated. Many pairs of variables that are of importance to behavioral scientists tend to be linearly related; and although there are many other ways of describing relationships, the linear model is the simplest, and the one most frequently used in such fields as psychology, education, and sociology.

DESCRIBING THE LINEAR RELATIONSHIP BETWEEN TWO VARIABLES

Suppose you want to measure the degree of linear relationship between a scholastic aptitude test (SAT) and college grade-point average (GPA). It would be reasonable to expect these two variables to be positively correlated, so that students with high

SAT scores, on the average, obtain relatively high GPAs and students with low SAT scores, on the average, tend to obtain low GPAs. The phrase " on the average " alerts us to the fact that there will be exceptions ; some with low SAT scores will do well in college and obtain high GPAs, while some with high SAT scores will do poorly and receive low GPAs. That is, the relationship between SAT scores and GPA is not perfect. Thus, the question arises : just how strong *is* the relationship ? How can it be summarized in a single number ?

THE Z SCORE DIFFERENCE FORMULA FOR r

We have seen in Chapter 6 that in order to compare scores on different variables (such as a mathematics test and a psychology test), it is useful to transform the raw scores into standard scores. These transformed scores allow you to compare paired values directly. Similarly, in order to obtain a coefficient of relationship which describes the similarity between paired measures in a single number, the raw score must be transformed into Z units where

$$Z_X = \frac{X - \bar{X}}{\sigma_X}$$

and

$$Z_Y = \frac{Y - \bar{Y}}{\sigma_Y}$$

In Table 11.1, a distribution of SAT scores (X) and GPAs (Y) is presented along with the corresponding descriptive statistics for

Figure 11.1 *Relationship between income of a family and IQ of a child in that family.*

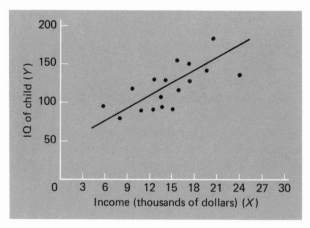

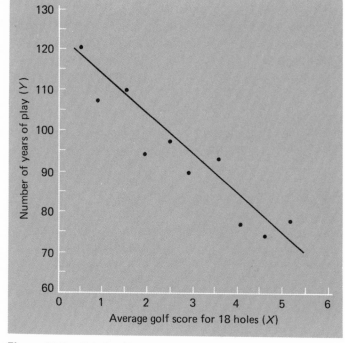

Figure 11.2   *Relationship between years of play and average score for ten golfers.*

Figure 11.3   *Relationship between length of big toe and IQ scores for adult men.*

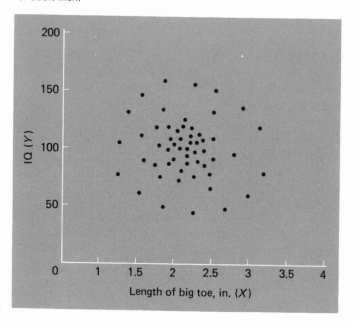

Table 11.1   *Raw scores and Z scores on SAT (X) and GPA (Y) for 25 students in an Eastern U.S. college*

student	X	Y	Z_X	Z_Y
1	650	3.8	1.29	1.67
2	625	3.6	1.08	1.31
3	480	2.8	−.16	−.09
4	440	2.6	−.50	−.44
5	600	3.7	.86	1.49
6	220	1.2	−2.37	−2.89
7	640	2.2	1.21	−1.14
8	725	3.0	1.93	.26
9	520	3.1	.18	.44
10	480	3.0	−.16	.26
11	370	2.8	−1.09	−.09
12	320	2.7	−1.52	−.26
13	425	2.6	−.62	−.44
14	475	2.6	−.20	−.44
15	490	3.1	−.07	.44
16	620	3.8	1.04	1.67
17	340	2.4	−1.35	−.79
18	420	2.9	−.67	.09
19	480	2.8	−.16	−.09
20	530	3.2	.27	.61
21	680	3.2	1.55	.61
22	420	2.4	−.67	−.79
23	490	2.8	−.07	−.09
24	500	1.9	.01	−1.67
25	520	3.0	.18	.26

$\bar{X} = 498.4$ $\bar{Y} = 2.8$ $\sigma_X = 117.33$ $\sigma_Y = .57$

25 students in a college in the eastern United States. The SAT scores were obtained when the students were high school seniors, and the GPAs were those received by the students after one year of college. In addition to the raw scores, the Z equivalents are also shown in the table. Notice that students with high SAT scores do tend to have high GPAs, and consequently large Z_X and Z_Y values; students with low SAT scores tend to have low GPAs. Thus, the paired Z values are similar for most students, and the two variables are therefore highly positively correlated. The *raw scores*, however, need not be numerically similar for any pair since they are calculated in different units for each variable; the pairs of Z scores more readily convey the precise relationship for any one individual.

If the association between the two selected variables were perfect and in a positive direction, each person would have exactly the same Z_X and Z_Y paired values. If the relationship were perfect but in a negative direction, each Z_X value would be paired with an identical Z_Y value but they would be *opposite in sign*. (See Figure 11.4.) These perfect relationships are offered for illustration and almost never occur in practice.

Since the size of the difference between paired Z values is related to the amount of the relationship between the two variables, some kind of *average* of these differences should yield information about the closeness or *strength* of the association between the variables. Because the mean of the differences $Z_X - Z_Y$ is necessarily zero, the coefficient of correlation is actually obtained by *squaring* differences between paired Z values. The size of the average of these squared differences,

$$\frac{\sum (Z_X - Z_Y)^2}{N}$$

is an index of the strength of the relationship. For example, a small value indicates a high positive correlation (little difference between the paired Z values). If this average is a large number, it indicates a high negative relationship (most paired Z values "opposite" to one another). A "medium size" average indicates little or no relationship. This average, however, is not convenient to interpret, since (as can be proved) it ranges from zero in the case of a perfect positive correlation to 4.0 in the case of a perfect

Figure 11.4 *Perfect linear relationships between two variables. (A) Perfect positive linear relationship; (B) perfect negative linear relationship.*

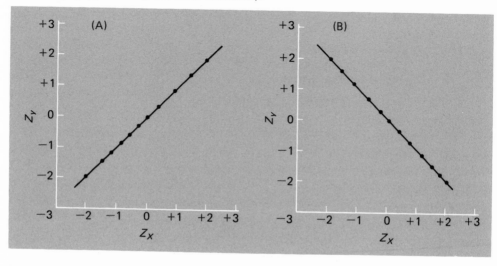

negative relationship. A much more readily interpretable coefficient is obtained by subtracting half of this average from one:

$$r_{XY} = 1 - \frac{1}{2} \frac{\sum (Z_X - Z_Y)^2}{N}$$

where r_{XY} = symbol for the correlation coefficient between X and Y
N = number of *pairs*

This correlation coefficient has the following desirable characteristics:

1. A value of zero indicates no linear relationship between the two variables (that is, they are linearly uncorrelated).
2. The *size* of the numerical value of the coefficient indicates the *strength* of the relationship (large absolute values mean that the two variables are closely related, and small absolute values mean that they are only weakly related).
3. The *sign* of the coefficient indicates the *direction* of the relationship.
4. The largest possible positive value is +1.00, and the largest possible negative value is −1.00.

Thus, if the correlation between two variables is +.20, you can tell at a glance that the relationship between them is positive and weak (since it is far from the maximum of +1). A correlation of −.80 would indicate a strong negative relationship.

The correlation coefficient may be symbolized more simply by r when there is no possible confusion as to which two variables are involved. This coefficient is often referred to as the Pearson r in honor of Karl Pearson, who did the early work on this measure starting with an idea of Francis Galton.

To clarify the meaning of the above formula, let us suppose that we have a small set of four paired X and Y scores, obtained from two quizzes in a small college seminar. (Four pairs would be far too few to permit a useful conclusion in a real experiment, but it will be easier to illustrate the concept if very few numbers are involved.) Let us first suppose that the results are as follows:

student	X	Y	Z_X	Z_Y	$Z_X - Z_Y$
1	8	29	+1.266	+1.266	0
2	7	26	+.633	+.633	0
3	5	20	−.633	−.633	0
4	4	17	−1.266	−1.266	0

Note that the paired Z values are identical in each case, indicating a perfect positive relationship. There is no difference between any of the corresponding Z values, so the correlation coefficient is equal to

$$r = 1 - \frac{1}{2} \frac{(0)^2 + (0)^2 + (0)^2 + (0)^2}{4}$$

$$= 1 - \tfrac{1}{2}(0)$$

$$= +1.00$$

Thus, the smallest possible mean of $(Z_X - Z_Y)^2$ is 0 and therefore the largest possible positive value of r is $+1$, the value that occurs when the relationship between the two variables is perfect and positive.

Now let us suppose that the results are instead as follows:

student	X	Y	Z_X	Z_Y	$Z_X - Z_Y$
1	8	17	+1.266	−1.266	+2.532
2	7	20	+.633	−.633	+1.266
3	5	26	−.633	+.633	−1.266
4	4	29	−1.266	+1.266	−2.532

In this case, the relationship between the two variables is perfect and negative, as is indicated by the fact that the Z values for each person are equal but opposite in sign. In other words, high scores on X are paired with low scores on Y, and low scores on X are paired with high scores on Y. The correlation coefficient is equal to:

$$r = 1 - \frac{1}{2} \frac{(+2.532)^2 + (+1.266)^2 + (-1.266)^2 + (-2.532)^2}{4}$$

$$= 1 - \frac{1}{2} \frac{16}{4}$$

$$= 1 - \tfrac{1}{2}(4)$$

$$= 1 - 2$$

$$= -1.00$$

Thus, the largest possible value of the mean of $(Z_X - Z_Y)^2$ is 4, as stated above, and therefore the largest possible negative value of r is -1, the value that occurs when the relationship between the two variables is perfect and negative.

Finally, let us suppose that the results are instead as follows:

student	X	Y	Z_X	Z_Y	$Z_X - Z_Y$
1	8	20	+1.266	−.633	+1.899
2	7	29	+.633	+1.266	−.633
3	5	17	−.633	−1.266	+.633
4	4	26	−1.266	+.633	−1.899

Here, there is no linear relationship at all between X and Y, as is shown by the paired Z values, and the correlation coefficient is equal to

$$r = 1 - \frac{1}{2} \frac{(+1.899)^2 + (-.633)^2 + (+.633)^2 + (-1.899)^2}{4}$$

$$= 1 - \frac{1}{2} \frac{8}{4}$$

$$= 1 - \tfrac{1}{2}(2)$$

$$= 1 - 1$$

$$= 0$$

The correlation coefficient can take on any value between -1.00 and $+1.00$. Equal numerical values of r describe equally strong relationships between variables; for example, coefficients of $+.50$ and $-.50$ describe relationships which are equally strong but are in opposite directions.

COMPUTING FORMULAS FOR r

Unfortunately, this procedure for obtaining r, although best for understanding the meaning of r, is much too tedious computationally even if desk calculators are available. It can be shown,* however, that identical results are obtained by calculating the mean of the *product* of the paired Z values:

$$r_{XY} = \frac{\sum Z_X Z_Y}{N}$$

The preceding formula can also be written as:

$$r_{XY} = \frac{\sum (X - \bar{X})(Y - \bar{Y})}{N \sigma_X \sigma_Y}$$

which, when written out, gives the *raw score formula for the Pearson correlation coefficient*:

$$r_{XY} = \frac{N \sum XY - \sum X \sum Y}{\sqrt{[N \sum X^2 - (\sum X)^2][N \sum Y^2 - (\sum Y)^2]}}$$

While the raw score formula looks formidable, it is in fact the easiest to use in practice. Note that the formula calls for both

---

* Proofs of the equivalence of all of the various formulas for r are given in the Appendix at the end of this chapter.

$\sum XY$ and $\sum X \sum Y$, which are *not* the same (see Chapter 1), and that N in all of the above formulas stands for the number of *pairs* of observations, that is, the number of cases.

The calculation of the Pearson r for the SAT and GPA data in Table 11.1 by both the raw score and Z product formula is shown in Table 11.2. The obtained value of $+.65$ indicates that for this group of 25 students, there is a high positive correlation between these two variables. (We call this value "high" because larger values for r between different variables are rare in the fields where r is used.) The corresponding scatter plot is shown in Figure 11.5.

CORRELATION AND CAUSATION

Some cautions must be noted at this point. First, you cannot determine the *cause* of the relationship from the correlation coefficient. Two variables may be highly correlated for one of three reasons: (1) X causes Y, (2) Y causes X, or (3) both X and Y are caused by some third variable. A well-known story that illustrates the danger of inferring causation from a correlation coefficient deals with a study which reported a high positive correlation between the number of storks and the number of births in European cities (that is, the more storks, the more births). Instead of issuing a dramatic announcement supporting the mythical powers of

Figure 11.5   *Scatter plot for data in Table 11.1* ($r = +.65$).

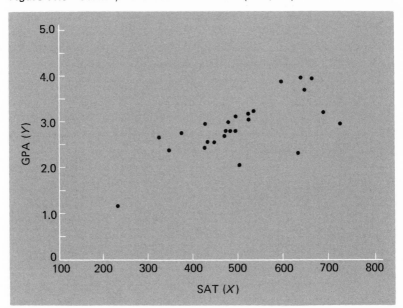

Table 11.2 *Calculation of Pearson correlation coefficient between SAT (X) and GPA (Y) by the raw score and Z product methods*

	raw score method					Z product method		
subject	X	Y	XY	X^2	Y^2	Z_X	Z_Y	$Z_X Z_Y$
1	650	3.8	2470	422500	14.44	1.29	1.67	2.1543
2	625	3.6	2250	390625	12.96	1.08	1.31	1.4148
3	480	2.8	1344	230400	7.84	−.16	−.09	.0144
4	440	2.6	1144	193600	6.76	−.50	−.44	.2200
5	600	3.7	2220	360000	13.69	.86	1.49	1.2814
6	220	1.2	264	48400	1.44	−2.37	−2.89	6.8493
7	640	2.2	1408	409600	4.84	1.21	−1.14	−1.3794
8	725	3.0	2175	525625	9.00	1.93	.26	.5018
9	520	3.1	1612	270400	9.61	.18	.44	.0792
10	480	3.0	1440	230400	9.00	−.16	.26	−.0416
11	370	2.8	1036	136900	7.84	−1.09	−.09	.0981
12	320	2.7	864	102400	7.29	−1.52	−.26	.3952
13	425	2.6	1105	180625	6.76	−.62	−.44	.2728
14	475	2.6	1235	225625	6.76	−.20	−.44	.0880
15	490	3.1	1519	240100	9.61	−.07	.44	−.0308
16	620	3.8	2356	384400	14.44	1.04	1.67	1.7368
17	340	2.4	816	115600	5.76	−1.35	−.79	1.0665
18	420	2.9	1218	176400	8.41	−.67	.09	−.0603
19	480	2.8	1344	230400	7.84	−.16	−.09	.0144
20	530	3.2	1696	280900	10.24	.27	.61	.1647
21	680	3.2	2176	462400	10.24	1.55	.61	.9455
22	420	2.4	1008	176400	5.76	−.67	−.79	.5293
23	490	2.8	1372	240100	7.84	−.07	−.09	.0063
24	500	1.9	950	250000	3.61	.01	−1.67	−.0167
25	520	3.0	1560	270400	9.00	.18	.26	.0468

$\sum X = 12{,}460$

$\sum Y = 71.2$

$\sum XY = 36{,}582$

$\sum X^2 = 6{,}554{,}200$

$\sum Y^2 = 210.98$

$(\sum X)^2 = (12{,}460)^2$
$= 155{,}251{,}600$

$(\sum Y)^2 = (71.2)^2$
$= 5069.44$

$N = 25$

$\sum Z_X Z_Y = 16.35$

$N = 25$

$r = \dfrac{\sum Z_X Z_Y}{N} = \dfrac{16.35}{25} = +.65$

$$r = \frac{N \sum XY - \sum X \sum Y}{\sqrt{[N \sum X^2 - (\sum X)^2][N \sum Y^2 - (\sum Y)^2]}}$$

$$r = \frac{(25)(36{,}582) - (12{,}460)(71.2)}{\sqrt{[25(6{,}554{,}200) - 155{,}251{,}600][25(210.98) - 5069.44]}}$$

$$r = +.65$$

storks, further investigation was carried out. It was found that storks nest in chimneys, which in turn led to the conclusion that a third variable was responsible for the relationship between storks and births—size of city. Large cities had more people, and hence more births; and more houses, and hence more chimneys, and hence more storks. Smaller cities had fewer people, births, houses, chimneys, and storks. Thus, attributing causality is a logical or scientific problem, not a statistical one.

CORRELATION AND RESTRICTION OF RANGE

A second important point has to do with the effect of the variability of the scores on the correlation coefficient. Suppose you calculate the correlation between achievement test scores and elementary school grades for public school children in New York City. You will probably find a strong linear trend when the data are plotted. But what if you were to calculate the correlation between the same two variables for a group of children in a class for the highly gifted? Here, the range of scores on both variables is considerably narrowed (or *attenuated*), as is shown in Figure 11.6. Because of this, the correlation between the two variables is markedly reduced. In other words, it is much harder to make fine discriminations among cases that are nearly equal on the variables in question than it is to make discriminations among cases that differ widely; it is difficult to predict whether a gifted child will be an A or A— student, but it is more feasible to distinguish among

Figure 11.6 *Effect of restriction of range on the correlation coefficient.*

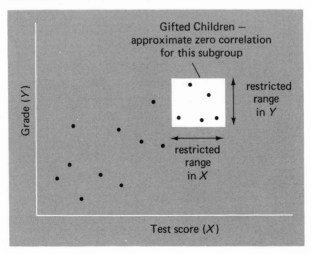

a broad range of A to F students. Thus, *when the range of scores is restricted* on either or both of two variables, the correlation between them decreases in absolute value.

TESTING THE SIGNIFICANCE OF THE CORRELATION COEFFICIENT

The correlation coefficient of +.65 for the group of 25 students whose scores are shown in Table 11.1 is a single number that conveniently describes the linear relationship between SAT scores and GPA *for this group.* It would be very useful to know, however, whether these two variables are correlated in the population of all students; that is, to draw an inference about likely values of the population correlation coefficient (symbolized by rho, ρ). Is it likely that there is actually no correlation in the population and that the correlation in the sample of 25 students was due to sampling error (the cases that happened to wind up in the sample)? Or, is the value of .65 large enough for us to conclude that there is a positive correlation between SAT scores and GPA in the population?

The strategy for testing hypotheses about likely values of ρ is similar to that used in previous chapters. The null hypothesis most often tested is that ρ is equal to zero*:

H_0: $\rho = 0$
H_1: $\rho \neq 0$

A criterion of significance, such as the .05 or .01 criterion, is selected, and the appropriate t ratio can then be computed,† with degrees of freedom equal to $N - 2$. Then, H_0 is retained if the computed t is less than the critical value of t from the t table; otherwise H_0 is rejected in favor of H_1 and r is said to be *significantly different from zero* (or, simply, *statistically significant*).

However, you do not necessarily need to compute the t ratio. This has (in a sense) already been done for you by statisticians who have constructed tables of significant values of r. Therefore, the procedure for testing a correlation coefficient for statistical

---

* Other null hypotheses are possible, but a different statistical procedure is required in order to test them. See Hays, W. L., *Statistics for psychologists*. New York: Holt, 1963. Pp. 529–532.

† The formula for the t ratio is

$$t = \frac{r\sqrt{N-2}}{\sqrt{1-r^2}}$$

where N = number of *pairs* of scores.

significance can be quite simple: you compare your computed value of r to the value of r shown in Table D in the Appendix for $N - 2$ degrees of freedom, where N is equal to the number of pairs. If the absolute value of your computed r is smaller than the tabled value, retain H_0; otherwise, reject H_0.

As an illustration, suppose that you wish to test the significance of the correlation between SAT scores and GPA, using $a = .05$. Referring to Table D, you find that for $25 - 2$ or 23 degrees of freedom, the smallest absolute value of r that is statistically significant is .396. Since the obtained r of .65 exceeds this value, you reject H_0 and conclude that there *is* a positive correlation in the population from which the sample of 25 students was selected. (Note that a correlation of $-.65$ would indicate a statistically significant negative relationship; the sign of r is ignored when comparing the computed r to the tabled value of r.)

IMPLICATIONS OF RETAINING H_0

If you fail to reject H_0, you have *not* established that the two variables are linearly uncorrelated in the population. Unless you know the probability of a Type II error (see Chapter 13), you cannot tell to what extent your decision to retain H_0 is likely to be wrong, and you should therefore limit yourself to a conservative statement such as "there is not sufficient reason to believe that ρ is different from zero."

Also, the Pearson r detects only linear relationships. Therefore, the possibility remains that the two variables are related, but not in a linear fashion. It is quite possible for two variables to be even perfectly related yet to have r equal to zero. In Figure 11.7, for example, you can predict Y without error given X, but *not* linearly.

IMPLICATIONS OF REJECTING H_0

If a Pearson r is statistically significant, the significance denotes *some* degree of linear relationship between the two variables in the population. It does *not* denote a "significantly high" or "significantly strong" relationship; it just means that the relationship in the population is unlikely to be zero. Notice that the larger the sample size, the smaller the absolute value of the correlation coefficient needed for statistical significance. For example, for $N = 12$ ($df = 10$), a correlation of .576 or larger is needed for significance using the .05 criterion, while for $N = 102$ ($df = 100$), a correlation of only .195 or larger is needed. This implies that the

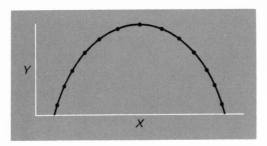

Figure 11.7 *Example of perfect relationship between two variables where r = 0.*

importance of obtaining statistical significance can easily be exaggerated. For example, suppose that for a very large sample (say, $N = 1000$), a statistically significant Pearson r of .08 is obtained. Although the statistical test indicates that ρ is greater than zero, the obtained r of .08 is so close to zero that the relationship in the population, although not zero and not necessarily equal to .08, is likely to be in the vicinity of .08 and therefore *very weak*—too weak, in fact, to allow you to make any accurate statements about one variable based on knowledge of the other. Consequently, such a finding would add little or nothing to our immediate knowledge of the environment even though it is statistically significant (although it may be useful in appraising a theory). Therefore, you need to have *both* statistical significance *and* a reasonably high absolute value of r before ρ is likely to be not only larger than zero, but large enough to indicate a sufficiently close relationship to be useful in applied work.

Unfortunately, correlation coefficients *cannot* simply be interpreted as percents; for example, you *cannot* conclude that a correlation of .60 is 60% of a perfect relationship or that it is twice as much as a correlation of .30. As we will see later in this chapter, however, the *squared* correlation coefficient (r^2) does permit an interpretation, in percentage terms, of the strength of the relationship between the two variables.

As usual, given that H_0 is true, the probability of erroneously rejecting it (a Type I Error) is equal to the criterion of significance that is selected. That is, using $a = .05$, you will reject H_0 5% of the time when it is actually true.

ASSUMPTIONS UNDERLYING THE USE OF r

The most important assumption underlying the use of the Pearson r is that X and Y are linearly related. If, for example, X and Y have a curvilinear relationship (as in Figure 11.7), the Pearson r will not detect it. Even this is not, strictly speaking, an assumption; if r is considered a measure of the degree of *linear* relationship, it remains such a measure whether or not the best-fitting function is linear. Ordinarily, however, one would not be interested in the best linear fit when it is known that the relationship is not linear.

Otherwise, no assumptions are made at all in using r to *describe* the degree of linear relationship between two variables for a given set of data. When testing a correlation coefficient for statistical significance, it is, strictly speaking, assumed that the underlying distribution is the so-called bivariate normal—that is, scores on Y are normally distributed for each value of X, and scores on X are normally distributed for each value of Y. However, when the degrees of freedom are greater than 25 or 30, failure to meet this assumption has little consequence to the validity of the test.

PREDICTION AND LINEAR REGRESSION

Behavioral scientists are indebted to Sir Francis Galton for making explicit some elementary concepts of relationship and prediction. When Galton wrote his now classic paper in 1885, *Regression toward mediocrity in hereditary stature*, he presented the theory that the physical characteristics of offspring tend to be related to, but are on the average less extreme than, those of their parents. According to his theory, for example, tall parents on the average produce children *less tall* than themselves, and short parents on the average produce children *less short* than themselves. In other words, physical characteristics of offspring tend to "regress" toward the average of the population. Thus, if you were to predict the height of a child from a knowledge of the height of the parents, you should predict a less extreme height—one closer to the average of all children.

Plotting data on the stature of many pairs of parents and off-spring, Galton calculated the median height of offspring for each "height" category of parents. For example, he plotted the median height for all offspring whose fathers were 5 ft 7 in., the median height for all offspring whose fathers were 5 ft 8 in., and so forth. By connecting these points representing the medians, Galton found not only that there was a positive relationship between parental height and height of the offspring, but also that this relationship was fairly linear. The line connecting each of the medians (and after Pearson, the means) came to be known as the *regression line*. This term has been adopted by statisticians (without its biological implications) to indicate the line used in predicting values of one variable from a knowledge of values of another variable with which it is paired.

Once the correlation between two variables is determined, the coefficient can be used to compute the equation for a regression line to obtain *predicted Y* values (symbolized by Y') from X, and the equation for another regression line to obtain *predicted X* values (denoted by X') from Y. (The two lines will be identical only if $r = \pm 1$.) The way in which a regression line is used to obtain a predicted score is illustrated in Figure 11.8. Suppose a person has a score of 20 on X and you want to predict what his score on Y will be. Given the regression line for predicting Y shown in Figure 11.8A, you enter the X axis at 20 and proceed up to the regression line. You then read out the predicted Y score; Y' for this subject is equal to 25. If, on the other hand, you want to predict the X score of a person whose score on Y is 32 and the regression line for predicting X is as shown in Figure 11.8B, you enter the Y axis at 32 and proceed across to the regression line. You then read out the predicted X score; X' for this subject is equal to 24. Before discussing the procedures used to calculate these regression lines, let us first consider the principles of prediction that are involved.

PRINCIPLES OF LINEAR REGRESSION

The regression (or prediction) line is the straight line which best represents the trend of the dots in a scatter plot. In any real

Figure 11.8 *Use of regression lines to obtain predicted scores.*

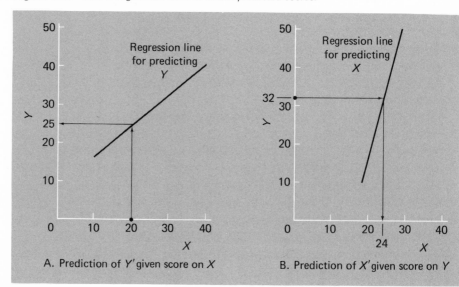

A. Prediction of Y' given score on X

B. Prediction of X' given score on Y

situation, the dots will *not* all fall exactly on a straight line. Therefore, you will make *errors* when you use a regression line to make predictions. The error in predicting a particular Y value is defined as the difference (keeping the sign) between the predicted Y value (Y') and the actual Y value (Y). That is,

error in predicting $Y = Y' - Y$

Similarly, the error in predicting X is the predicted X' value minus the actual X value, that is, $X' - X$. The two regression lines are compared in Figure 11.9 for a small set of data, and the concept of errors in prediction is illustrated separately for predicting Y (Figure 11.10) and predicting X (Figure 11.11). In Figure 11.10, for example, note that the predicted Y value for the individual with an X score of 68 is equal to approximately 159.8. His actual Y value, however, is equal to 148. Thus, the difference of approximately $+11.8$ between the predicted and actual values represents an error in prediction for this individual; the positive sign indicates that it is an *over*-prediction. Similar reasoning applies to the other cases in Figure 11.10.

The lower the absolute value of the correlation between X and Y (that is, the closer it is to zero), the greater the absolute size of the errors, or disparities between the predicted and actual values, will be. For example, the correlation between SAT scores and GPA was found to be .65, so a predicted GPA (Y') is likely to be close to the actual GPA (Y) obtained by the student. On the other hand, the correlation between length of big toe and IQ is approximately .00, so you are likely to predict values of IQ (Y') that are quite different from the actual IQ (Y) of the individual in this instance. In fact, it can readily be shown that the absolute value of $r_{XY} = r_{YY'}$—that is, the absolute value of the correlation between X and Y is equal to the correlation between the actual Y values and the predicted Y values obtained from the regression line for predicting Y. Also, the absolute value of $r_{XY} = r_{XX'}$. Only if $r_{XY} = \pm1$ will the "errors" all be zero.

You may be wondering why we are concerned with predicted Y scores in Figure 11.10. After all, we know the Y values (weights) for all of the cases shown in Figure 11.10 and do not need to predict them. The answer is that *you can use the regression line to make predictions for new cases for whom you have data only on the X variable.* Thus, the first step in a regression problem is to collect data on *both* the X and Y variables. For example, to predict college grade point average from scores on the SAT, you must first obtain a sample of college students who have scores on both the SAT and GPA. Using this sample, you compute the regression line. You can then use the regression line to predict GPA for high-school seniors who have scores only on the SAT.

The regression line actually used is the one which minimizes the sum of the *squared* errors in prediction. That is, the chosen regression line for predicting Y gives a value for

$$\sum (Y' - Y)^2$$

which is smaller than one would obtain using any other line with which to make the prediction for these data. It is therefore called the *least squares* regression line of Y on X. Similarly, the chosen regression line for predicting X minimizes $\sum (X' - X)^2$, and is the least squares line for X on Y.

COMPUTING THE EQUATIONS FOR PREDICTING Y FROM X AND X FROM Y*

You may recall from high school algebra that the equation for any straight line takes the form

$$Y' = bX + a$$

Figure 11.9 *Regression lines when height scores (X) are plotted against weight scores (Y) for ten adult males.*

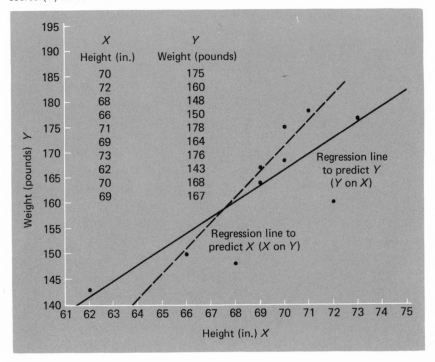

* Although in the two-variable case it is often difficult to label the dependent and independent variables, the variable being predicted is conventionally called the *dependent* or *criterion* variable (often symbolized by Y) and the second variable is called the *independent* or *predictor* variable (often symbolized by X).

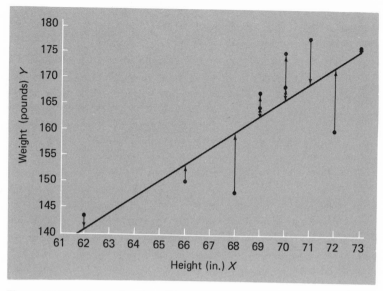

Figure 11.10   *Regression line of Y on X showing extent of error (difference between actual weight score and predicted weight score).**

Figure 11.11   *Regression line of X on Y showing extent of error (difference between actual height score and predicted height score).†*

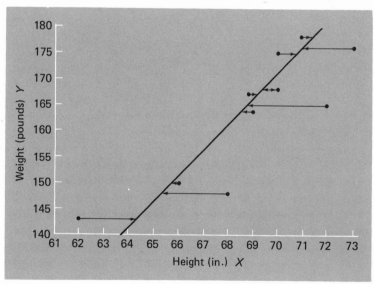

* Vertical lines are drawn between each observed weight score and the predicted weight score for each value of X (actual height). These vertical distances are the errors in prediction. All *predicted* scores lie on the straight line. Note that there are *two* height scores of 70 in. with two different weight scores, and two height scores of 69 in. with two different weight scores; the smaller error is superimposed upon the larger in each of these cases.

† Horizontal lines indicate the extent of error between predicted height and actual height for each value of Y (actual weight).

In this equation, a is a constant equal to the value of Y which corresponds to an X value of zero; that is, it is the point at which the straight line crosses the Y axis. This value is therefore called the Y *intercept*. The other constant, b, indicates the rate at which Y changes with changes in X. That is, if you were to select any two values of X (denoted by X_1 and X_2) with their corresponding Y values (Y_1 and Y_2) and plot these two values, the value of b would be equal to:

$$b = \frac{Y_2 - Y_1}{X_2 - X_1}$$

In regression analysis, the b coefficient is written as b_{YX} when predicting Y from X, and it is written as b_{XY} to indicate prediction of X from Y. The b value is also the tangent of the angle made by the line with the horizontal axis and is called the *slope* of the line.

As indicated in the preceding section, the objective when predicting Y is to obtain the regression line which minimizes the value of $\sum (Y' - Y)^2$. It can be shown, using calculus, that the values of a and b which accomplish this are:

$$b_{YX} = r_{XY} \frac{\sigma_Y}{\sigma_X}$$

$$a = \bar{Y} - b_{YX}\bar{X}$$

Rewriting the equation for a straight line with these values substituted for a and b gives

$$Y' = r_{XY} \frac{\sigma_Y}{\sigma_X} X + (\bar{Y} - b_{YX}\bar{X})$$

or

$$Y' = r_{XY} \frac{\sigma_Y}{\sigma_X} (X - \bar{X}) + \bar{Y} \qquad \text{regression equation to predict } Y \text{ values}$$

When the goal is to predict X, the objective is to minimize $\sum (X' - X)^2$. Using calculus, it can be shown that the values of a and b which accomplish this are

$$b_{XY} = r_{XY} \frac{\sigma_X}{\sigma_Y}$$

$$a = \bar{X} - b_{XY}\bar{Y}$$

Note that $r_{XY} = r_{YX}$ but b_{XY} (used for predicting X) is *not* equal to b_{YX} (used for predicting Y). Substituting the above values into the equation for a straight line gives

$$X' = r_{XY} \frac{\sigma_X}{\sigma_Y} Y + (\bar{X} - b_{XY}\bar{Y})$$

or

$$X' = r_{XY} \frac{\sigma_X}{\sigma_Y} (Y - \bar{Y}) + \bar{X} \qquad \text{regression equation to predict } X \text{ values}$$

If you are working with raw scores and r is not available, it may prove easier to calculate the value of b_{YX} or b_{XY} from the following computing formulas:

For predicting Y:

$$b_{YX} = \frac{N \sum XY - \sum X \sum Y}{N \sum X^2 - (\sum X)^2}$$

For predicting X:

$$b_{XY} = \frac{N \sum XY - \sum X \sum Y}{N \sum Y^2 - (\sum Y)^2}$$

An illustration of the use of the prediction equation for Y is given in Table 11.3. For each X score, the *predicted* Y score (Y') has been

Table 11.3 *Use of the regression equation $Y' = .0032X + 1.26$ to predict Y scores (GPA) from X scores (SAT)*

subject	X	predicted Y (Y')	actual Y	error (Y' − Y)
1	650	.0032(650) + 1.26 = 3.3	3.8	−0.5
20	530	.0032(530) + 1.26 = 3.0	3.2	−0.2
3, 19	480	.0032(480) + 1.26 = 2.8	2.8	0.0
22	420	.0032(420) + 1.26 = 2.6	2.4	+0.2

obtained by use of the regression equation to predict Y values shown above. We have already seen that $\bar{X} = 498.4$, $\bar{Y} = 2.8$, $\sigma_X = 117.33$, $\sigma_Y = .57$, and $r_{XY} = .65$ (see Tables 11.1 and 11.2). Substituting these values in the regression equation gives

$$Y' = (.65) \frac{.57}{117.33} (X - 498.4) + 2.8$$

$$= (.65) \frac{.57}{117.33} X - (.65) \frac{.57}{117.33} (498.4) + 2.8$$

$$= .0032X + 1.26$$

This equation is used to obtain predicted grade-point average (Y') given a SAT score (X). For example, the predicted grade-point average for student 1 in Table 11.3 who has a SAT score of 650 is obtained by substituting 650 for X:

$$Y' = (.0032)(650) + 1.26$$
$$= 3.3$$

Thus, for students who receive a score of 650 on the SAT, you would predict that their grade-point average at the end of one year in college would be 3.3. This is the *best* linear prediction you can make, the one which on the average would be least in error.

The procedure shown above is repeated for each X value shown in Table 11.3, and the results are listed in the Y' column. The actual values of Y and the amount of error in each case are also shown. As an exercise, you may wish to calculate the Y' values for all 25 students whose scores are given in Table 11.1 and compute the correlation between the actual and the predicted values of Y (that is, $r_{YY'}$). You should obtain a value equal to that of r_{XY} or .65.*

The regression equation can also be written in terms of Z scores, where the predicted Z_Y value (symbolized by $Z_{Y'}$) is equal to $(Y' - \bar{Y})/\sigma_Y$ and Z_X is equal to $(X - \bar{X})/\sigma_X$. If in the equation for Y' we move $\bar{Y}$ to the other side of the equal sign and then divide both sides of the equation by σ_Y and rearrange the terms on the right side of the equal sign, it becomes apparent that

$$Z_{Y'} = r_{XY}Z_X$$

Thus, the *predicted* Y score expressed as a Z value is equal to the X score expressed as a Z value multiplied by the correlation coefficient. This equation may prove helpful in understanding linear regression. For example, it is evident that the predicted score $(Z_{Y'})$ will be less extreme (that is, closer to its mean) than the score from which the prediction is made (Z_X) because Z_X is multiplied by a fraction (r_{XY}). The equation also shows that if $r_{XY} = 0$, all predicted $Z_{Y'}$ values will equal zero, which is equivalent to the mean of the Y' scores $(\bar{Y})$. Thus, when X gives no accurate information as to likely values of Y, the best you can do is make the "safe" prediction of an average Y score for everyone. This incidentally shows that the mean is a least squares measure—that is, the sum of the squared deviations of the values in the sample from it is a minimum.

Although this equation looks simple, it is not in general convenient for calculating predicted scores. To use it for this purpose, you would have to carry out the following steps: (1) transform X

---

* Since Y' is a linear transformation of X, $r_{XY'} = 1$. If the correlation between any two variables equals 1, they have equal correlations with any other variable (such as Y). In this vein, note that the correlation between X and Z_X or any other linear transformation of X is equal to 1 (or -1 if the slope of the transformation is negative). This implies that applying a linear transformation to either or both variables does *not* change the absolute value of r.

to a Z value; (2) compute the value of $Z_{Y'}$; (3) transform $Z_{Y'}$ to a raw score equivalent Y'. Most of the time, it will be easier to use the raw score regression equation.

MEASURING PREDICTION ERROR: THE STANDARD ERROR OF ESTIMATE

We have seen that when predicting scores on Y, the amount of squared error made for a given individual is equal to $(Y' - Y)^2$. The *average* squared error for the entire sample can be obtained by summing the squared errors and dividing by N. Such a measure of error, however, would be in terms of squared units. As was the case with the standard deviation (Chapter 5), a measure in terms of actual score units can be obtained by taking the positive square root. This gives a useful measure of prediction error which is called the *standard error of estimate*, symbolized by $\sigma_{Y'}$:

$$\sigma_{Y'} = \sqrt{\frac{\sum (Y' - Y)^2}{N}}$$

In practice, the standard error of estimate can more easily be obtained by the following formula, which can be proved to be equivalent to the one above by algebraic manipulation:

$$\sigma_{Y'} = \sigma_Y \sqrt{1 - r_{XY}^2}$$

Note that if $r = \pm 1$, there is no error; while if $r = 0$, the standard error of estimate reaches its maximum possible value σ_Y.

The standard error of estimate of Y may be thought of as the variability of Y about the regression line *for a particular X value, averaged over all values of X*. This is illustrated in Figure 11.12. In Figure 11.12A, the variability of the Y scores within each X value is low, since the scores within each X value cluster closely together. The total variability of Y (for all X combined), which is given by σ_Y is fairly large. Thus, $\sigma_{Y'}$ is *small compared to* σ_Y. This is good for purposes of prediction; an individual with a low score on X is likely to have a low score on Y, while an individual with a high score on X is likely to have a high score on Y. Thus, in this instance, Y is defined within fairly narrow limits if X is known. In Figure 11.12B, however, prediction will be very poor. The average variability within each X value is about equal to the total variability of Y; that is, $\sigma_{Y'}$ is *about equal to* σ_Y. Therefore, in this instance, knowing an individual's X score will *not* permit you to make a good prediction as to what his Y score will be. This illustration shows that

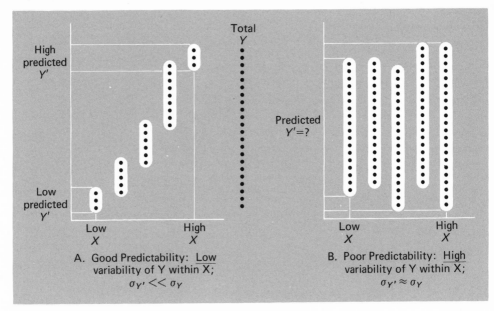

Figure 11.12 *Relationship between the standard error of estimate* ($\sigma_{Y'}$), *the standard deviation of Y* (σ_Y), *and the accuracy of prediction.*

prediction will be good to the extent that *knowing X reduces variability in predicting Y.*

Note that in Figure 11.12A, *X* and *Y* are highly linearly correlated; while in Figure 11.12B, the correlation is approximately zero. The standard error of estimate is also helpful in interpreting the meaning of a given numerical value of *r*, for it can be shown by algebraic manipulation that

$$r_{XY}^2 = 1 - \frac{\sigma_{Y'}^2}{\sigma_Y^2}$$

$$= \frac{\sigma_Y^2 - \sigma_{Y'}^2}{\sigma_Y^2}$$

That is, the *squared* correlation coefficient shows *by what proportion the reduction in the variability (specifically the variance) in predicting or accounting for Y has been achieved by knowing X.* If this reduction has been great (as in Figure 11.12A), $\sigma_{Y'}^2$ is small compared to σ_Y^2 and r^2 is large. If, on the other hand, little reduction has been achieved (as in Figure 11.12B), $\sigma_{Y'}^2$ is about equal to σ_Y^2 and r^2 is about equal to zero. In the example involving GPA and SAT scores, r^2 is equal to $(.65)^2$ or .42. Thus, in this example, *42% of the variance of Y is explained by variance in X,* so a fairly good job has been done of identifying a (linear) relationship between two important variables.

When predicting X scores from Y, the standard error of estimate (denoted by $\sigma_{X'}$) is:

$$\sigma_{X'} = \sqrt{\frac{\sum (X' - X)^2}{N}}$$

or

$$\sigma_{X'} = \sigma_X \sqrt{1 - r_{XY}^2}$$

The use and underlying logic of $\sigma_{X'}$ is similar to $\sigma_{Y'}$.

SUMMARY

Correlation refers to the co-relationship between two variables. The Pearson r coefficient is a useful measure of *linear* correlation; the *sign* of r indicates the *direction* of the relationship and the *size of the numerical value of r* indicates the *strength* of the relationship. The maximum values of the Pearson r are -1 (perfect negative linear relationship) and $+1$ (perfect positive linear relationship). The Pearson r can be tested for statistical significance; it *cannot*, however, be interpreted as a percent (but r^2 can).

Linear regression is used to predict scores on one variable given scores on the other. The *standard error of estimate* is a measure of error in the prediction process.

APPENDIX TO CHAPTER 11: EQUIVALENCE OF THE VARIOUS FORMULAS FOR r

1.
$$1 - \frac{1}{2} \frac{\sum (Z_X - Z_Y)^2}{N} = \frac{\sum Z_X Z_Y}{N}$$

PROOF Expanding the term in parentheses gives

$$r = 1 - \frac{1}{2} \frac{\sum (Z_X^2 - 2Z_X Z_Y + Z_Y^2)}{N}$$

$$= 1 - \frac{1}{2} \frac{\sum Z_X^2 - 2 \sum Z_X Z_Y + \sum Z_Y^2}{N} \qquad \text{(Rules 1, 2, and 8, Chapter 1)}$$

$$= 1 - \frac{1}{2} \left(\frac{\sum Z_X^2}{N} - \frac{2 \sum Z_X Z_Y}{N} + \frac{\sum Z_Y^2}{N} \right)$$

It can readily be shown that the sum of squared Z scores divided by N is equal to 1:

$$\sigma = \frac{\sum (Z - \bar{Z})^2}{N} \qquad \text{(definition of } \sigma^2\text{)}$$

$$1 = \frac{\sum (Z - 0)^2}{N} \qquad \text{(mean of } Z \text{ scores} = 0; \text{ variance of } Z \text{ scores} = 1\text{)}$$

$$1 = \frac{\sum Z^2}{N}$$

Thus,

$$r = 1 - \frac{1}{2}\left(1 - \frac{2\sum Z_X Z_Y}{N} + 1\right)$$

$$= 1 - 1 + \frac{\sum Z_X Z_Y}{N}$$

$$= \frac{\sum Z_X Z_Y}{N}$$

Incidentally, we know that the correlation of Z_X (or any other variable) with *itself* is $+1$, so $(\sum Z_X Z_X)/N = +1$. But this is the same as $(\sum Z_X^2)/N = 1$, which we have already seen.

2. $$\frac{\sum Z_X Z_Y}{N} = \frac{\sum (X - \bar{X})(Y - \bar{Y})}{N\sigma_X \sigma_Y}$$

PROOF   By definition,

$$Z_X = \frac{X - \bar{X}}{\sigma_X}, \qquad Z_Y = \frac{Y - \bar{Y}}{\sigma_Y}$$

Therefore,

$$r = \frac{\sum \left(\frac{X - \bar{X}}{\sigma_X}\right)\left(\frac{Y - \bar{Y}}{\sigma_Y}\right)}{N}$$

$$= \frac{\sum (X - \bar{X})(Y - \bar{Y})}{N\sigma_X \sigma_Y} \qquad \text{(Rule 8, Chapter 1)}$$

3.   Prove that

$$\frac{\sum (X - \bar{X})(Y - \bar{Y})}{N\sigma_X \sigma_Y}$$

and hence the other formulas in this appendix, are equal to the raw score computing formula.

PROOF We will deal with the numerator and denominator separately. Expanding the numerator gives

$$\sum (X - \bar{X})(Y - \bar{Y}) = \sum (XY - X\bar{Y} - \bar{X}Y + \bar{X}\bar{Y})$$
$$= \sum XY - \bar{Y} \sum X - \bar{X} \sum Y + N\bar{X}\bar{Y}$$

(Rules 1, 2, 5, and 8, Chapter 1)

Substituting $(\sum X)/N$ for $\bar{X}$ and $(\sum Y)/N$ for $\bar{Y}$ gives

$$\sum XY - \frac{(\sum Y)(\sum X)}{N} - \frac{(\sum X)(\sum Y)}{N} + \frac{N(\sum X)(\sum Y)}{(N)(N)}$$
$$= \sum XY - 2\frac{(\sum X)(\sum Y)}{N} + \frac{(\sum X)(\sum Y)}{N}$$
$$= \sum XY - \frac{\sum X \sum Y}{N}$$

Turning to the denominator, we have seen in Chapter 5 that

$$\sigma_X = \sqrt{\frac{1}{N}\left[\sum X^2 - \frac{(\sum X)^2}{N}\right]}$$

$$\sigma_Y = \sqrt{\frac{1}{N}\left[\sum Y^2 - \frac{(\sum Y)^2}{N}\right]}$$

Therefore,

$$N\sigma_X \sigma_Y = N \sqrt{\frac{1}{N}\left[\sum X^2 - \frac{(\sum X)^2}{N}\right]} \sqrt{\frac{1}{N}\left[\sum Y^2 - \frac{(\sum Y)^2}{N}\right]}$$

Since $[N(\sqrt{1/N})(\sqrt{1/N})]$ equals N/N or 1, these terms cancel out, leaving

$$\sqrt{\left[\sum X^2 - \frac{(\sum X)^2}{N}\right]\left[\sum Y^2 - \frac{(\sum Y)^2}{N}\right]}$$

If we now combine the numerator and the denominator, we have

$$r = \frac{\sum XY - \dfrac{\sum X \sum Y}{N}}{\sqrt{\left[\sum X^2 - \dfrac{(\sum X)^2}{N}\right]\left[\sum Y^2 - \dfrac{(\sum Y)^2}{N}\right]}}$$

Finally, if we multiply this by $N/\sqrt{N^2}$, which is equal to 1, we have

$$r = \frac{N \sum XY - \sum X \sum Y}{\sqrt{[N \sum X^2 - (\sum X)^2][N \sum Y^2 - (\sum Y)^2]}}$$

which is the raw score computing formula for r.

12 other correlational techniques

THE RELATIONSHIP BETWEEN RANKED VARIABLES: THE SPEARMAN RANK-ORDER CORRELATION COEFFICIENT

The Pearson correlation coefficient r is an appropriate statistic for describing the relationship between two variables when a linear model is appropriate and when data are continuous. For other kinds of data, however, other correlational techniques are necessary. For example, suppose two psychologists *rank* a group of 10 patients in terms of degree of pathology, where a rank of 1 indicates the least degree of pathology and a rank of 10 indicates the greatest degree. The results are shown in Table 12.1. Did the two psychologists rank the patients in a similar manner?

Ranks provide less information than do continuous scores. For example, in Table 12.1, it is evident that Observer X considered

Table 12.1 *Ten patients ranked for degree of pathology by two independent observers (1 = least, 10 = most)*

patient's first name	ranks given by observer X	ranks given by observer Y	D (X − Y)	D² (X − Y)²
John	3	4	−1	1
Joseph	2	1	1	1
Charles	5	6	−1	1
William	9	7	2	4
Robert	1	3	−2	4
Bruce	10	10	0	0
Jennings	8	9	−1	1
Daniel	4	2	2	4
Richard	7	5	2	4
Raymond	6	8	−2	4
				$\sum D^2 = 24$

Robert as the least pathological and Joseph as the next lowest in pathology. With continuous scores such as test scores, you could see at a glance whether these two patients are similar on this characteristic (for example, Robert scores 72 on a personality test and Joseph scores 74) or different (for example, Robert scores 72 and Joseph scores 95). With ranks, however, there is no indication of the *distance* between these two patients; Observer X might have felt that Robert was just slightly less pathological than Joseph *or* he might have felt that there was a considerable difference between them. The statistic most frequently used to describe and to test hypotheses about the relationship between ranks is the Spearman rank-order correlation coefficint, symbolized as r_s,* and it is computed as follows:

$$r_s = 1 - \frac{6 \sum D^2}{N(N^2 - 1)}$$

where $D =$ difference between a pair of ranks

$N =$ number of pairs

If you instead applied the formula for the Pearson product moment correlation coefficient (given in Chapter 11) to rank-order data by treating the ranks as scores and there were no tied ranks, you would get exactly the same value as r_s; the formula for r_s is simply a short-cut for the formula for r that takes advantage of certain relationships which hold when the N scores are the integers from 1 to N. Thus, r_s varies between -1.0 and $+1.0$, and a high positive value of r_s indicates a strong tendency for the paired ranks to be equal, while a high negative value of r_s indicates a strong tendency for the paired ranks to be k and $N - k$ (for example, patients ranked as most pathological by Observer X would be ranked as least pathological by Observer Y, an outcome which would undoubtedly produce some bewilderment on the part of the psychologists). A zero value would indicate no relationship between the two sets of ranks.

In our example, the calculation of $\sum D^2$ is shown in Table 12.1. First, the difference (D) between each pair of ranks is computed, where $D = X - Y$; then, each difference is squared; finally, the

---

* The Greek letter rho, ρ, is also commonly used to designate the rank-order correlation coefficient. However, in order to avoid confusion between this statistic and the population parameter for the Pearson correlation coefficient, which is also symbolized as rho, we will use r_s for the Spearman coefficient.

squared differences are summed. Then, r_s is equal to

$$r_s = 1 - \frac{6(24)}{10(10^2 - 1)}$$

$$= 1 - \frac{144}{990}$$

$$= 1 - .15$$

$$= .85$$

TIED RANKS

In the preceding example, Observer X might have decided that Joseph and John were about equal in pathology and that both should therefore share second place. This would be expressed numerically by *averaging* the two ranks in question and assigning this average as the rank of each person. The ranks in question are 2 and 3, and the average of these ranks, or 2.5, is the rank given to both John and Joseph. The next patient would receive a rank of 4. What if three patients tied for the fourth rank? In this case, ranks 4, 5, and 6 would be averaged and each of these three patients would receive a rank of 5. The next patient would then be ranked 7th.

When there are tied ranks, the above formula for r_s will give a result which overestimates the absolute value of r_s. However, unless there are many ties, and particulary long ties (that is, three- or more-way ties), this overestimate is likely to be trivial, say less than .01 or .02. If you want to be safe, abandon the r_s formula and apply the regular Pearson product-moment r formula (Chapter 11) to the ranks as if they were scores.

TESTING THE SIGNIFICANCE OF THE RANK-ORDER CORRELATION COEFFICIENT

Although using either the formula for r_s or the formula for the Pearson r will yield the same numerical result in the absence of tied ranks, the significance of the two coefficients *cannot* be tested in the same manner when N is very small (less than 10). To test the null hypothesis that the ranks are independent in the population from which the sample was drawn, you can use Table E in the Appendix. The minimum values of r_s needed for statistical significance are shown in the table for each N. (Note that, when using this table, you need only refer to N—the number of pairs of ranks—rather than degrees of freedom.)

For example, to use Table E for the problem involving the data in Table 12.1, you would look up the minimum value necessary for r_s to be statistically significant when $N = 10$. Using the .05 criterion, this value is .648, and since the absolute value of the obtained r_s (.85) is greater than the tabled value, you reject H_0 and conclude that there is a statistically significant relationship between the two sets of ranks.

Note that when N is greater than or equal to 10, the procedures for testing a Pearson r for statistical significance will give a very good approximation. That is, you can compute

$$t = \frac{r_s\sqrt{N-2}}{\sqrt{1-r_s^2}}$$

and refer the obtained t to the t table with $N - 2$ degrees of freedom, rejecting H_0 in favor of H_1 if the absolute value of the computed t exceeds (or equals) the tabled value. Or, you can refer the computed correlation to Table D in the Appendix with $N - 2$ degrees of freedom.

This test of signifance is not, in general, as *powerful* as tests of significance using score data (for example, the test for the Pearson r). That is, it is *not* as likely to detect true significant relationships. Therefore, it would *not* be a good idea to convert continuous data into rank-order data in order to use this statistical test. It is not unusual, however, for data to exist only in rank-order form, and such instances are the *raison d'etre* of the Spearman rank-order correlation coefficient.

THE RELATIONSHIP BETWEEN ONE DICHOTOMOUS AND ONE CONTINUOUS VARIABLE

THE POINT BISERIAL CORRELATION COEFFICIENT

Suppose you are interested in the relationship between two variables, and one of them is continuous and the other is genuinely *dichotomous* (only two categories exist on the variable, such as male—female, presently married—presently unmarried, yes—no). You can compute the correlation between them by letting any number (say 0) stand for one of the dichotomized categories and any other number (say 1) for the other and using the usual formula for the Pearson r (see Chapter 11). Thus, the two numbers (for example, the 0 and 1) are treated just as any other scores. It does not matter which two numbers you use because, as was pointed out in Chapter 11, the Pearson *r* *standardizes* each variable; whatever numbers you use, the Z scores will not change in absolute value.

This correlation coefficient is called the *point biserial correlation coefficient*, symbolized as r_{pb}. While you can use the computational procedures discussed in Chapter 11 if you wish, there is an easier way to compute r_{pb}:

$$r_{pb} = \frac{N \sum Y_1 - N_1 \sum Y}{\sqrt{N_1 N_0 [N \sum Y^2 - (\sum Y)^2]}}$$

Let the dichotomized variable (which equals 1 or 0) be symbolized by X. Then:

Y = continuous variable

$\sum Y_1$ = sum of Y values for those observations for which the associated X value = 1

$\sum Y$ = sum of *all* the Y values

$\sum Y^2$ = sum of all squared Y values

N_1 = the number of observations where $X = 1$

N_0 = the number of observations where $X = 0$

N = the total number of paired observations

$\quad = N_1 + N_0$

As an illustration, consider the data in Table 12.2, where the goal is to determine the relationship between sex and arithmetic achievement scores for a random sample of high school seniors. Arbitrarily, it is decided to score males as 1 on the dichotomous variable and females as 0, and r_{pb} is obtained as follows:

$$\sum Y_1 = 10 + 15 + 15 + 20 + 5 + 10 + 10 + 5 = 90$$
$$\sum Y = 350$$
$$N_1 = 8$$
$$N_0 = 12$$
$$N = 20$$
$$\sum Y^2 = 7800$$
$$(\sum Y)^2 = (350)^2 = 122{,}500$$

$$r_{pb} = \frac{(20)(90) - (8)(350)}{\sqrt{(8)(12)[(20)(7800) - (122{,}500)]}}$$

$$= \frac{1800 - 2800}{\sqrt{(96)[156{,}000 - 122{,}500]}}$$

$$= -.56$$

Table 12.2   *Scores on arithmetic achievement test for 20 high school boys and girls*

sex (X)[a]	achievement scores (Y)
1	10
1	15
0	30
0	20
0	25
1	15
0	20
0	25
0	30
1	20
1	5
0	5
1	10
0	10
0	20
1	10
0	30
0	35
1	5
0	10
$\sum X = 8$	$\sum Y = 350$
	$\sum Y^2 = 7800$

[a] Male $= 1$, female $= 0$.

Thus, there is a high negative relationship between sex and achievement scores. Since males have been assigned a score of 1 and females a score of 0, this means that there is a negative relationship between arithmetic achievement and being male or a positive relationship between arithmetic achievement and being female. If you had assigned females a score of 1 and males a score of 0, you would have obtained an r_{pb} of $+.56$.

The point biserial correlation coefficient is tested for statistical significance in the same way as the Pearson r. Thus, comparing the obtained value of $-.56$ to the critical value of .444 obtained from Table D in the Appendix for $N - 2$ or 18 df, you would decide to reject H_0 in favor of H_1 because the absolute value of the obtained r is greater than the tabled value. You would therefore conclude that a negative correlation between arithmetic achieve-ment and being male exists in the population from which this sample of 20 paired observations was drawn.

CONVERTING t VALUES FOR THE DIFFERENCE BETWEEN TWO MEANS TO r_{pb} VALUES

An alternative way of testing the significance of the data in Table 12.2 would be to form two groups, males and females, and apply the procedures of Chapter 10 to the data:

males	females
10	30
15	20
15	25
20	20
5	25
10	30
10	5
5	10
	20
	30
	35
	10

That is, you would compute

$$t = \frac{\bar{X}_{males} - \bar{X}_{females}}{\sqrt{s^2_{pooled}\left(\frac{1}{N_1}+\frac{1}{N_2}\right)}}$$

This procedure would yield exactly the same value of t you would get if you substituted the r_{pb} of $-.56$ in the formula for the t test of a Pearson correlation coefficient in Chapter 11. The point biserial correlation coefficient, however, has the advantage of providing an indication of the *size of the effect* as well as indicating statistical significance. In Chapter 11, we pointed out that a small Pearson r such as .08, even if statistically significant, would be al-most useless for practical purposes because the size of the relationship between the two variables in the population, while likely to be greater than zero, is also likely to be extremely small. This consideration is true of all procedures in inferential statistics—statistical significance does *not* necessarily imply a *large* relationship or effect; it just indicates that the effect in the population is *unlikely to be zero*. A t value does *not* offer a solution to this problem, since it does not provide a ready measure of how large the relationship is, that is, the *effect size*. Therefore, it is very desirable to convert significant values of t obtained from procedures such as those of Chapter 10 into a point biserial correlation coefficient and report *both* t (and statistical significance) and r_{pb} or r_{pb}^2, the proportion of variance in the continuous variable accounted for by membership in one group versus the other. It can be shown that this conversion can readily be accomplished by the following formula:

$$r_{pb} = \sqrt{\frac{t^2}{t^2 + df}}$$

When dealing with the t test for the difference between two means, df (as usual) is equal to $N_1 + N_2 - 2$.

As an illustration, consider once again the caffeine experiment discussed in Chapter 10. It was found that caffeine did have a positive effect on mathematics test scores; a statistically significant value of t was obtained. The question remains, however, as to *how strong* is the relationship between the presence or absence of that dosage of caffeine and test scores. Recalling that $t = 2.91$ and $df = (22 + 20 - 2) = 40$, we have

$$r_{pb} = \sqrt{\frac{(2.91)^2}{(2.91)^2 + 40}}$$

$$= \sqrt{\frac{8.47}{84.47}}$$

$$= \sqrt{.175}$$

$$= .42$$

Thus, there is a moderately strong relationship between the presence or absence of caffeine and test scores. In fact, r_{pb}^2 can be interpreted in the same way as r^2, so $(.42)^2$ or 17% of the variance in test scores is explained by whether or not caffeine is present. You therefore know that caffeine is a fairly important factor that accounts for differences among people on this test, though far from the only one—83% of the variance in test scores is still not accounted for. Assuming that caffeine has no known harmful side effects, a cautious recommendation that students take caffeine before a mathematics examination would be justified. If instead the statistically significant t had yielded a value of r_{pb} of .10, however, you know that only $(.10)^2$ or 1% of the variance in test scores is explained by the presence or absence of caffeine. You are therefore warned that this experimental finding, while statistically significant, is probably not of much practical value; and you certainly would *not* recommend that caffeine actually be used, for the probable effect on test scores would be too small to matter. The possibility does exist, however, that while such a finding is small and not of much practical import, it might be of considerable theoretical importance in understanding (for example) brain functioning.

Unfortunately, it is very common to report values of t in the professional literature *without* also reporting the corresponding value of r_{pb}. Thus, there exist many instances wherein researchers have exaggerated the importance of findings which are statistically significant but (unknown to them) are also so weak as to be almost useless in practice. We therefore suggest that you compute r_{pb} yourself when reading about t values in professional articles and books. You may well find that the researcher's enthusiastic commentary about a statistically significant result is in fact unjustified.*

THE BISERIAL CORRELATION COEFFICIENT

Suppose that a sample of 50 job applicants take an ability test. All are then hired; six months later, each employee's foreman rates him as either a "success" or a "failure." (The foremen do not feel able to give more precise ratings.) The question of interest is whether or not test scores, a continuous variable, are correlated with job success, a variable for which there are only two scores. Job success, however, is *not* a true dichotomy; there exists a continuum of job ability along which the employees differ. That is, unlike a genuine dichotomy such as male—female, the job success variable in this situation is actually a continuous variable on which only dichotomized scores are available.

In such cases, assuming that the variable which appears in crude dichotomous form is actually normally distributed and it is this underlying variable whose correlation we want to estimate, it is recommended that a different correlation coefficient (called the *biserial* correlation coefficient) be computed instead of r_{pb}. A

---

* For a more extensive discussion of converting t and other statistics into correlational-type indices, see: Cohen, J., Some statistical issues in psychological research. In Wolman, B. B. (Ed.) *Handbook of clinical psychology.* New York: McGraw-Hill, 1965.

discussion of the biserial correlation coefficient, however, is beyond the scope of this book; the interested reader is referred to Guilford.*

THE RELATIONSHIP BETWEEN TWO DICHOTOMOUS VARIABLES

THE PHI COEFFICIENT

When you wish to determine the relationship between two variables and *both* of them are genuinely dichotomous, the appropriate correlation coefficient to compute is the *phi coefficient*. This coefficient is closely related to the chi square statistic, and will therefore be discussed in the chapter dealing with chi square (Chapter 16).

THE TETRACHORIC CORRELATION COEFFICIENT

In some situations, you may wish to determine the relationship between two variables where both are actually continuous yet where only dichotomized scores are available on each one. As an illustration, suppose that 50 job applicants are interviewed; after each interview, the interviewer recommends either that the company hire the applicant or that they do not hire him. Regardless of the interviewer's rating, all 50 applicants are hired. Six months later, each employee's foreman rates him as either a "success" or a "failure." The company wishes to know the degree of relationship between the continuous, assumed normal variable which underlies the interviewer's recommendations and the similarly continuous normal variable underlying job success. Note that it is not the correlation between the observed dichotomies which is at issue—that would be given by the phi coefficient—but rather the relationship between the underlying variables.

In situations such as this, it is recommended that a coefficient called the *tetrachoric correlation* be computed. Since a discussion of the tetrachoric correlation is beyond the scope of this book, we will again refer the interested reader to Guilford.†

---

* Guilford, J. P., *Fundamental statistics in psychology and education* (4th Ed.). New York: McGraw-Hill, 1965. Pp. 317–321.
† Guilford, J. P., *Fundamental statistics in psychology and education* (4th Ed.). New York: McGraw-Hill, 1965. Pp. 326–332.

SUMMARY

The Spearman rank-order correlation coefficient, a Pearson r, is used when data are in the form of *ranks*. The point-biserial correlation coefficient, also a Pearson r, is used when one variable is *dichotomous* (has only two possible values) and one is continuous. It is particularly desirable to convert a statistically significant t value obtained from the difference between two means (Chapter 10) to a value of r_{pb} or r_{pb}^2 so as to obtain a measure of the *strength* of the relationship between the independent and dependent variables.

13 introduction to power analysis

When you perform a statistical test of a null hypothesis, the probability of a Type I error (that is, rejecting H_0 when it is actually true) is equal to the significance criterion α (see Chapter 9). Since a small value of α is customarily chosen (for example, .05 or .01), you can be confident *when you reject H_0* that your decision is very likely to be correct. In hypothesis testing, however, you also run a risk that even if H_0 is false you may fail to reject it— that is, you may commit a Type II error. The probability of a Type II error is β; and the complement of the probability of a Type II error, or $1 - \beta$, is the *probability of getting a significant result* and is called the *power* of the statistical test. Techniques for computing β or $1 - \beta$ were not discussed in Chapter 9.

A little contemplation immediately suggests that since a behavioral (or any other) scientist who tests a null hypothesis almost certainly wants to reject it, he will desire that the power of the statistical test be high rather than low. That is, he hopes that there is a good chance that the statistical test will indicate that he should switch to H_1 when it is correct to do so. Despite this, it is a strange fact that this topic has received little stress in the statistical textbooks used by behavioral scientists. The unfortunate result of this is that research is often done in which, unknown to the investigator, power is low (and, therefore, the probability of a Type II error is high), a false null hypothesis is not rejected (that is, a Type II error is actually made), and much time and effort are wasted. Worse, a promising line of research may be prematurely and mistakenly abandoned because the investigator does not know that he should have relatively little confidence concerning his failure to reject H_0.

This chapter will deal first with the concepts involved in power analysis. Then, methods will be presented for accomplishing the two major kinds of power analysis, which can be applied to the null hypothesis tests discussed thus far.

CONCEPTS OF POWER ANALYSIS

The material in this section, since it holds generally for a variety of statistical tests, is abstract and requires careful reading. In ensuing sections, these ideas will be incorporated into concrete procedures for power and sample size analysis of four major kinds of hypothesis tests.

There are four major parameters involved in power analysis:

1. The significance criterion, α. This, of course, is the familiar criterion for rejecting the null hypothesis and equals the probability of a Type I error, frequently .05. A little thought should convince you that the more stringent (the smaller) this criterion, other things being equal, the harder it is to reject H_0 and therefore the lower is the power. It also follows that, again other things including power being equal, a more stringent significance criterion requires a larger sample size for significance to be obtained.

2. The sample size, N. Whatever else the accuracy of a sample statistic may depend upon, it *always* depends on the size of the sample on which it has been determined. Thus, all the standard errors you have encountered in this book contain some function of N in the denominator. It follows that, other things being equal, error decreases and power increases as N increases.

3. The population "effect" size, γ **(gamma).** This is a new concept, and a very general one that applies to many statistical tests. When a null hypothesis about a population is false, it is false to some degree. (In other words, H_0 might be very wrong, or somewhat wrong, or only slightly wrong.) γ is a very general measure of the *degree* to which the null hypothesis is false, or how large the "effect" is in the population. Within the framework of hypothesis testing, γ can be looked upon as a *specific* value which is an alternative to H_0. Specific alternative hypotheses, in contrast to such universal alternative hypotheses as $H_1 : \mu_1 - \mu_2 \neq 0$ or $H_1 : \rho \neq 0$, are what make power analyses possible. We will consider γ in detail in subsequent sections.

Other things being equal, power increases as γ, the degree to which H_0 is false, increases. That is, you are less likely to fail to reject a false H_0 if H_0 is very wrong. Similarly, other things including power being equal, the larger the γ, the smaller the N which is required for significance to be obtained.

4. Power, or $1 - \beta$. The fourth parameter is power, the probability of rejecting H_0 for the given significance criterion. It is equal to the complement of the probability of a Type II error; that is, power $= 1 - \beta$.

These four parameters are mathematically related in such a way that any one of them is an exact function of the other three. We will deal with the two most useful ones:

1. Power determination. Given that a statistical test is performed with a specified α and N and that the population state of affairs is γ, the power of the statistical test can be determined.

2. N determination. Given that the population state of affairs is γ and a statistical test using α is to have some specified power (say .80), the necessary sample size, N, can be determined.

One final concept that will prove useful in statistical power analysis is δ (delta), where

$$\delta = \gamma f(N)$$

That is, δ is equal to γ times a function of N. Thus, δ combines the population effect size and the sample size into a single index. The table from which power is read (Table H in the Appendix) is entered using δ.

In the sections which follow, the general system described above for analyses of power determination and sample size determination will be implemented for four different statistical tests:

1. Tests of hypotheses about the mean of a single population (Chapter 9).
2. Tests of hypotheses about the proportion of a single population (Chapter 9).
3. Tests of the significance of a Pearson correlation coefficient (Chapter 11).
4. Tests of the difference between the means of two populations (Chapter 10).

The procedures to be described are approximate, and applicable for large samples (N at least 25 or 30). This is because the system uses the normal curve, which as we have seen is not only useful in its own right but is also a good approximation to the t distribution once N is that large.

THE TEST OF THE MEAN OF A SINGLE POPULATION

POWER DETERMINATION

In Chapter 9 we described how to test the null hypothesis that the mean height of midwestern American men is 68 in. The value of 68 in. was selected because it is the mean of U.S. adult males, and the issue is whether or not the midwestern men differ from that height. Thus, the null hypothesis is: $H_0: \mu = \mu_0 = 68$.

The alternative hypothesis stated in Chapter 9 was merely that $H_1: \mu \neq 68$. However, power analysis is impossible unless a *specific* H_1 is stated. Assume that you suspect (or are interested in the possibility) that the population mean of midwestern men is an inch away from 68: $H_1: \mu = \mu_1 = 67$ or 69. (This form is used to indicate a two-tailed test; if a one-tailed test were desired such that midwestern men were predicted as taller, H_1 would be $\mu = \mu_1 = 69$.) Assume further that a two-tailed .05 decision rule is to be used, and that the sample size is to be $N = 100$. (Assume also that you know or can estimate the population standard deviation, σ, of heights to be 3.1 in.) The power determination question can then be formulated as follows: If we perform a test at $a = .05$ of $H_0: \mu = 68$ using a random sample of $N = 100$, and in fact μ is 69 or 67, what is the probability that we will get a significant result and hence reject H_0?

The size of the effect postulated in the population, which is one inch in *raw score* terms, must be expressed as a γ value to accomplish the power analysis. For a test of the mean of a single population, γ is expressed essentially as a Z score:

$$\gamma = \frac{\mu_1 - \mu_0}{\sigma}$$

In the present example,

$$\gamma = \frac{69 - 68}{3.1} = \frac{1}{3.1} = .32$$

Note that this is *not* a statement about actual or prospective *sample* results, but expresses, as an alternative hypothesis, the *population* state of affairs. That is, μ_1 is postulated to be $.32\sigma$ away from μ_0, the value specified by H_0. Note also that the sign of γ is ignored, since the test is two-tailed.

Having obtained the measure of effect size, the next step is to obtain δ. In the previous section, we pointed out that $\delta = \gamma f(N)$.

For the test of the mean of a single population, the specific function of N is $\sqrt{N}$, so that

$$\delta = \gamma\sqrt{N}$$

In the present example,

$$\delta = .32\sqrt{100} = 3.2$$

Entering Table H in the Appendix with $\delta = 3.2$ and $a = .05$, the power is found to be .89. Thus, if the mean of the population of midwestern men is an inch away from 68, the probability of rejecting H_0 in this situation is .89 (or, the probability of a Type II error is .11). If the mean of the population of midwestern men is more than one inch away from 68, the power will be greater than .89, while if the population mean is less than one inch away from 68, the power will be less than .89.

It is important to understand that power analysis proceeds completely with population values and not sample results, either actual or prospective. The above analysis could well take place before the data were gathered to determine what the power would be under the specified a, γ, and N. Or it could take place after an experiment was completed to determine the power the statistical test *had*, given a, γ, and N with no reference to the obtained data. If the power for a reasonable postulated γ were equal to .25 when the results of the experiment were not statistically significant, the nonsignificant result would be inconclusive, since the *a priori* probability of obtaining significance is so small (and the probability of a Type II error is so high). On the other hand, power of .90 with a nonsignificant result tends to suggest that the actual γ is not likely to be as large as postulated.

In order to compute γ in the above example, it was necessary to posit a specific value of μ_1 (69 or 67) and also of σ (3.1). This can sometimes be difficult when the unit is not a familiar one, as for example when extensive a priori data are not available for a new test. For situations like this, it is useful to specify conventional values corresponding to "small," "medium," and "large" values of γ, which although arbitrary are reasonable (in much the same way as the .05 decision rule). For the test of the mean of a single population, these are:

small: $\gamma = .20$
medium: $\gamma = .50$
large: $\gamma = .80$

Thus, *if* you are unable to posit specific values of μ_1 or σ, you can select the value of γ corresponding to how large you believe the effect size in the population to be. Do *not*, however, use the above conventional values if you can determine γ values that are appropriate to the specific problem or field of research in which the statistical test occurs, for conventional values are only reasonable approximations. As you get to know a substantive research area, the need for reliance on these conventions should diminish.

SAMPLE SIZE DETERMINATION

We now turn to the other major kind of statistical power analysis, sample size determination. Here, α, γ, and the desired power are specified, and you wish to determine the necessary sample size. This form of analysis is particularly useful in experimental planning, since it provides the only rational method for making the crucially important decision about the size of N.

First, we must consider the concept of "desired power." Your first inclination might be to set power very high, such as .99 or .999. But, as your intuition might suggest, the quest for near certainty is likely to result in the requirement of a very large N, usually far beyond the researcher's resources. This is similar to the drawback of choosing a very small significance criterion, such as .0001; while a Type I error is very unlikely with such a criterion, you are also very unlikely to obtain statistical significance unless the effect size and/or sample size is unrealistically large. Therefore, just as it is prudent to seek less than certainty in minimizing Type I errors by being content with a significance criterion such as .05 or .01, a prudent power value is also in order insofar as Type II errors are concerned.

Although a researcher is of course free to set any power value that makes sense to him, we suggest the value .80, which makes the probability of a Type II error equal to .20, when a conventional standard is desired. The reason that this suggested value is larger than the customary .05 value for the probability of a Type I error is that in most instances in science, it is seen as less desirable to mistakenly reject a true H_0 (which leads to false positive claims) than to mistakenly fail to reject H_0 (which leads only to failure to find something and no claim at all; for example, "the evidence is insufficient to warrant the conclusion that . . ."). Besides, if desired power were conventionally set at .95 (making $\beta = .05$), most studies would demand larger samples than most investigators could muster. We repeat, however, that the .80 value should be used only as a general standard, with any investigator quite free to make his own decision.

Let us return to the mean height of midwestern men, but now change the task. Instead of assuming that N is to be 100, suppose you are now interested in the following question: "If I test at $a = .05$ the null hypothesis that $\mu = 68$ when in fact $\mu = 69$ or 67 (and $\sigma = 3.1$), how large must N be for me to have a .80 probability of rejecting H_0 (that is, have power $= .80$)?" To answer this question, the first step is to obtain two values: γ, determined by the same methods as in the preceding section and equal to $(\mu_1 - \mu_0)/\sigma$ or .32, and the desired power, specified as .80. Next, δ is obtained by entering Table I in the Appendix in the row for desired power $= .80$ and the column for a (two-tailed) $= .05$; δ is found to be 2.80. Finally, for this test of the mean of a single population, N is found as follows:

$$N = \left(\frac{\delta}{\gamma}\right)^2$$

In the present example,

$$N = \left(\frac{2.80}{.32}\right)^2$$

$$= (8.75)^2$$

$$= 77$$

Thus, to have power $= .80$ in the present situation, the sample must have 77 cases in it. Note that this is consistent with the previous result where we saw that, other things (γ, a) equal, $N = 100$ resulted in power $= .89$.

To illustrate the consequences of demanding very high power, consider what happens in this problem if desired power is set at .999. From Table I in the Appendix, $\delta = 5.05$. Substituting,

$$N = \left(\frac{\delta}{\gamma}\right)^2 = \left(\frac{5.05}{.32}\right)^2 = (15.78)^2 = 249$$

Thus, in this problem, to go from .80 to .999 power requires increasing N from 77 to 249, which is more than many researchers can manage. Of course, if data are easily obtained or if the cost of making a Type II error is great, no objection can be raised about such a "maximum" power demand.

THE SIGNIFICANCE TEST OF THE PROPORTION OF A SINGLE POPULATION

POWER DETERMINATION

The structure of the system remains the same as before; only the details need to be adjusted. The null hypothesis in question is that the population proportion, π, is equal to some specified value. That is, $H_0: \pi = \pi_0$. H_0 is tested against some specific alternative, $H_1: \pi = \pi_1$. For example, let us return to the worried politician in Chapter 9 who wants to forecast the results of an upcoming two-man election by obtaining a random sample of $N = 400$ voters and testing the null hypothesis that the proportion favoring him is $\pi_0 = .50$. He thinks that he is separated from his opponent by about .08. That is, he expects the vote to be .54 to .46 or .46 to .54. (His expectations are stated in both directions because a two-tailed significance test is intended.) The question can be summarized as follows: If a statistical test is performed at $a = .05$ of $H_0: \pi = .50$ against the specific alternative $H_1: \pi = .54$ (or .46) with $N = 400$, what is the power of this statistical test?

For the test of a proportion from a single population, the effect size, γ, and δ are defined as follows:

$$\gamma = \frac{\pi_1 - \pi_0}{\sqrt{\pi_0(1 - \pi_0)}}$$

$$\delta = \gamma\sqrt{N}$$

For the data of this problem,

$$\gamma = \frac{.54 - .50}{\sqrt{.50(1 - .50)}}$$

$$= \frac{.04}{.50}$$

$$= .08$$

$$\delta = .08\sqrt{400}$$

$$= 1.60$$

Entering Table H with $\delta = 1.60$ and $a = .05$, the power is found to be .36. Thus, our worried politician has something else to worry about: as he has planned the study, he has only about one chance in three of coming to a positive conclusion (rejecting the null hypothesis that $\pi = .50$) if the race is as close as he thinks ($H_1: \pi = .54$ or .46). If the poll had already been conducted as described, and the result were not statistically significant, he should consider the results inconclusive; even if π is as far from .50 as .54 or .46, the probability of a Type II error (β) is .64. Note that even if the two-tailed test were to be performed at $a = .10$, the power would only be .48.

SAMPLE SIZE DETERMINATION

As in the preceding section, let us now invert the problem to one of determining N, given the same $\gamma = .08$, $a = .05$ and specifying the desired power as .80. The formula for N for the test of a proportion from a single population, like the above formula for δ, is the same as in the case of the mean of a single population:

$$N = \left(\frac{\delta}{\gamma}\right)^2$$

The value of δ for the joint specification of power $= .80$, α (two-tailed) $= .05$ is found from Table I to be 2.80. Therefore,

$$N = \left(\frac{2.80}{.08}\right)^2$$

$$= (35)^2$$

$$= 1225$$

This sample size is considerably larger than the $N = 400$ which yielded power of .36 under these conditions. It is hardly a coincidence that in surveys conducted both for political polling and market and advertising research, sample sizes typically run about 1500.

If conventional values for effect size for this test are needed, the following γ values can be used: small, .10; medium, .30; large, .50.

THE SIGNIFICANCE TEST OF A PEARSON r

POWER DETERMINATION

The term Pearson r is intended to be used inclusively for *all* such rs, that is, for the rank order, point-biserial, and phi coefficients as well as those computed from continuous interval scales. Power analysis of significance tests of the Pearson r of a sample proceed quite simply. Recall from Chapter 11 that for such tests, the null hypothesis is: $H_0: \rho = 0$. For the purpose of power analysis, the alternative hypothesis is that the population value is some *specific* value other than zero: $H_1: \rho = \rho_1$. For example, you might be testing an r for significance when expecting that the population value is .30, in which case $H_1: \rho = .30$. (Again, for two-tailed tests the value is taken as either $+.30$ or $-.30$.) The γ value for this test requires no formula; it is simply the value of ρ specified by H_1 (.30 in this example). To find δ, the appropriate function of N to be combined with γ is $\sqrt{N-1}$:

$$\delta = \gamma\sqrt{N-1} = \rho_1\sqrt{N-1}$$

For example, consider a study in which it is to be determined whether there is a linear relationship between a questionnaire measure of introversion and some physiological measure. A reasonably random sample of 50 Psychology 1 students at a large university are used for a two-tailed statistical test with $a = .01$. The researcher expects the population ρ to be .30; thus $\rho_1 = .30$ while H_0: $\rho = 0$. What is the power of the test?

We can go directly to

$$\delta = .30\sqrt{50 - 1}$$
$$= 2.10$$

Entering Table H for $\delta = 2.10$ and $a = .01$, power is found to be .32. A one chance in three of finding significance may strike the researcher as hardly worth the effort. He may then reconsider the stringency of his significance criterion, and check $a = .05$; the power of .56 with this more lenient criterion well may still not satisfy him. He might then plan to increase his sample size (see below).

If the investigator has difficulty in formulating an alternative-hypothetical value for ρ_1, the following conventional values are offered: small, .10; medium, .30; large, .50. The value of .50 may not seem "large," but over most of the range of behavioral science where correlation is used, correlations between different variables do not often get much larger than that.

SAMPLE SIZE DETERMINATION

Returning to the introversion study above, we can ask what N is necessary for a test at a (two-tailed) $= .05$ and assuming $\gamma = \rho_1 = .30$ in order to have power of (let us say) .75. The value of N is just one more than for the other two one-sample tests described above:

$$N = \left(\frac{\delta}{\gamma}\right)^2 + 1 = \left(\frac{\delta}{\rho_1}\right)^2 + 1$$

δ is found from Table I for power $= .75$, $a = .05$ to be 2.63, so

$$N = \left(\frac{2.63}{.30}\right)^2 + 1$$
$$= (8.77)^2 + 1$$
$$= 78$$

Thus, to have a .75 chance (that is, three to one odds) of finding r to be significant if $\rho_1 = +.30$ (or $-.30$), he needs a sample of 78 cases.

TESTING THE SIGNIFICANCE OF THE DIFFERENCE
BETWEEN INDEPENDENT MEANS

The final significance test whose power analysis we consider is the test of the difference between the means of two independently drawn random samples. This is probably the most frequently performed test in the behavioral sciences.

As stated in Chapter 10, the null hypothesis most frequently tested is H_0: $\mu_1 - \mu_2 = 0$ (often written as $\mu_1 = \mu_2$). For power analysis, a *specific* alternative hypothesis is needed, which we will write as H_1: $\mu_1 - \mu_2 = \theta$ (theta), where θ is the difference between the means expressed in *raw* units. To obtain γ, the standard measure of effect size, this difference must be standardized, and this is done using the standard deviation of the population, σ (which is a single value since we assume that $\sigma_1 = \sigma_2$ for the two populations). Thus, the value of γ in this instance is conceptually the same as for the test of a single population mean, namely

$$\gamma = \frac{\text{alternative-hypothetical } \mu_1 - \mu_2}{\sigma} = \frac{\theta}{\sigma}$$

This can be looked upon as the difference between Z score means of the two populations, or, equivalently, the difference in means expressed in units of σ. Again, the direction of the difference (the sign of θ) is ignored in two-tailed tests.

This device of "standardizing" the difference between two means is generally useful, and not only for purposes of power analysis. Frequently in behavioral science, there is no sure sense of how large a unit of raw score is. How large *is* a point? Well, they come in different sizes: An IQ point comes about 15 to the standard deviation (σ_{IQ}) while an SAT point is much smaller, coming 100 to the standard deviation (σ_{SAT}). By always using σ as the unit of measurement, we achieve comparability from one measure to another, as was the case with Z scores.

This device helps us out with regard to other related problems. One occurs whenever we have not had much or any experience with a measure (a new test, for example) and we have little if any basis for estimating the population σ, which the above formula for γ requires. Paradoxically enough, we can use this σ as our unit, despite the fact that it is unknown, by thinking directly in terms of γ. Thus, a γ of .25 indicates a difference between two population means (whose exact values we do not know) equal to .25σ (whose exact value we also do not know). It is as if our ignorance cancels out, and in a most useful sense $\gamma = .25$ is always the same size difference whether we are talking about IQ, height, socio-economic status, or a brand-new measure of "oedipal intensity."

Well, how large *is* a γ of .25, or any other? Here, as before, it is possible (and sometimes necessary) to appeal to some conventions, and they are the same as in the test of the mean of a single population: small, .20; medium, .50; large, .80. In this framework, a γ of .25, for example, could be characterized as a "smallish" difference. Again we point out that when an investigator has a firmer basis for formulating γ, these conventions should not be used. Frequently, however, they will come in handy.

POWER DETERMINATION

For the comparison of two means, the value of δ is equal to:

$$\delta = \gamma \sqrt{\frac{N}{2}}$$

where $N =$ size of *each* of the two samples (thus $2N$ cases are needed in all)

As an illustration, consider the caffeine experiment in Chapter 10 (slightly revised). Given two groups, each with $N = 40$ cases, one of which is given a dose of caffeine and the other a placebo, we will test at $a = .05$ (two-tailed) the difference between their means on a mathematics test following the treatment. If there is in the populations a mean difference of medium size, operationally defined as $\gamma = .50$, what is the power of the test? With γ given on the basis of convention, there is no need to compute it, and hence no need to estimate μ_1, μ_2, or σ. Thus, all we need is

$$\delta = .50 \sqrt{\frac{40}{2}}$$
$$= 2.24$$

Entering Table H in the column for $a = .05$, we find for $\delta = 2.2$ that power $= .59$ and for $\delta = 2.3$ that power $= .63$. Interpolating between these two values to obtain the power corresponding to $\delta = 2.24$ yields power of $.59 + .4(.63 - .59) = .61$. Thus, for medium differences (as defined here), samples of 40 cases each have only a .61 probability of rejecting H_0 for a two-tailed test using $a = .05$.

Ordinarily, in research where variables are to be manipulated such as the caffeine experiment described above, it is possible to arrange the experiment so that the total available pool of subjects is divided equally among the groups. This division is optimal. But it is not always possible or desirable to have equal sample sizes. For example, it would be a mistake because of loss of power to reduce the larger of two available samples to make it equal to the smaller. Instead, to determine power when $N_1 \neq N_2$, compute the so-called *harmonic mean* of the two Ns:

$$\text{harmonic mean} = \frac{2N_1N_2}{N_1 + N_2}$$

Then, use this obtained value for N in the formula for δ. As an example, consider a psychiatric hospital study where there are resources to place a random sample of 50 newly admitted schizophrenic patients into a special experimental treatment program and compare their progress after one month with a control group of 150 cases receiving regular treatment for the same period. It is planned to rate each patient on an improvement scale 30 days following admission and to test for the significance of the difference in means for rated improvement, using a two-tailed test where $a = .05$. If it is anticipated that $\gamma = .40$ (between "small" and "medium"), what is the power of the test?

Since $N_1 \neq N_2$, compute

$$\frac{2N_1N_2}{N_1 + N_2} = \frac{2(50)(150)}{50 + 150} = 75$$

Then enter this value as N in the formula for δ:

$$\delta = .40 \sqrt{\frac{75}{2}}$$
$$= 2.45$$

Interpolating between $\delta = 2.4$ and $\delta = 2.5$ in Table H in the $a = .05$ column yields power $= .69$.

Note that a total of $N_1 = N_2 = 200$ cases will be studied, but since $N_1 \neq N_2$, the resulting power is the same as if two samples of 75 cases each, or 150 cases, were to be studied; there is a smaller effective N and less power. This demonstrates the nonoptimality of unequal sample sizes. Were it possible to divide the total of 200 cases in two equal samples of 100, δ would equal $.40\sqrt{100/2}$ or 2.83, and the resulting power at $a = .05$ obtained from Table H by interpolation is .81.

SAMPLE SIZE DETERMINATION

The sample size needed for *each* independent sample when means are to be compared can be found for specified values of a, γ, and the desired power, as was shown for the other statistical tests in this chapter. The combination of a and power is expressed as δ by looking in Table I, and the values for δ and γ are substituted in the equation appropriate for this statistical test, which is

$$N = 2 \left(\frac{\delta}{\gamma} \right)^2$$

For example, it was found in the caffeine experiment that for $a = .05$ and $\gamma = .50$, the plan to use samples of $N = 40$ each resulted in power $= .61$. If power $= .90$ is desired, what N per sample is needed?

For a two-tailed test where $a = .05$ and power $= .90$, Table I yields $\delta = 3.24$. Substituting,

$$N = 2 \left(\frac{3.24}{.50} \right)^2$$

= 84 cases in each group, or 168 cases in all

Again we see that large desired power values demand large sample sizes. Were the more modest power of .80 specified, δ would equal 2.80 (Table I) and N would equal $2(2.80/.50)^2$ or 63, materially smaller number.

This chapter, as its title indicates, merely introduces the concepts of power analysis and presents some simple approximate procedures for performing it in connection with four statistical tests. For a more thorough but still esssentially nontechnical presentation, which covers all the major statistical tests with many tables which minimize the need for computation, see Cohen.*

SUMMARY

The probability of a Type II error (failing to reject a false H_0) and the corresponding probability of obtaining a significant result [the *power* of the statistical test, equal to $1 - P$ (Type II error)] are of major statistical importance but have, strangely enough, been virtually ignored in the behavioral sciences. If an experiment has a poor chance of yielding a significant result (that is, low power) it should be redesigned; negative findings will be inconclusive because the probability of a Type II error is high, and much time and effort is likely to be wasted. This chapter presents techniques for determining power which make possible the planning of efficient experiments designed to yield more conclusive statements if H_0 is not rejected.

* Cohen, J., *Statistical power analysis for the behavioral sciences*. New York: Academic Press, 1969.

14 one-way analysis of variance

In Chapter 10, techniques were given for testing the significance of the difference between two means. These techniques enable you to determine the effect of a single independent variable (for example, the presence or absence of a particular dosage of caffeine) on the mean of a dependent variable (for example, mathematics test scores) when there are *two* samples of interest, such as an experimental group and a control group.

Suppose that you would now like to determine whether *different dosages* of caffeine affect performance on a 20-item English test. As before, there is one independent variable (amount of caffeine), but this time you wish to include the following *five* samples:

Sample 1. very large dose of caffeine
Sample 2. large dose of caffeine
Sample 3. moderate dose of caffeine
Sample 4. small dose of caffeine
Sample 5. placebo (no caffeine)

As usual, you would like the probability of a Type I error in this experiment to be .05 or less. It would *not* be correct to perform ten separate *t* tests for the difference between two means (that is, first test $H_0: \mu_1 = \mu_2$; then test $H_0: \mu_1 = \mu_3$; and so on for the means of samples 1 versus 4, 1 versus 5, 2 versus 3, etc.) and compare each obtained *t* value to the appropriate critical *t* value for $a = .05$. The more statistical tests you perform, the more likely it is that some will be statistically significant purely by chance. That is, when $a = .05$, the probability that one *t* test will yield statistical significance when H_0 actually is true is .05; this is the probability of committing a Type I error. If you run 20 *t* tests when H_0 is always true, an average of one of them ($.05 \times 20 = 1.0$) will be statistically significant just on the basis of chance, so it is very likely that you will commit a Type I error somewhere along the line. Similarly, if you run ten separate *t* tests in order to test the null hypothesis that different dosages of caffeine do not affect English test scores, the probability that you will commit a Type I error is clearly *greater* than the desired .05.

A procedure for testing differences among three or more means for statistical significance which overcomes this difficulty is the *analysis of variance* (abbreviated as ANOVA). ANOVA can also be used with just two samples, in which case it yields identical results to those of the procedures given in Chapter 10. The null hypothesis tested by ANOVA is that the means of the populations from which the samples were randomly drawn are all equal; for example, the null hypothesis in the caffeine experiment is

$$H_0: \quad \mu_1 = \mu_2 = \mu_3 = \mu_4 = \mu_5$$

The alternative hypothesis states that H_0 is *not* true. It does not indicate anything about specific population means; for example, you cannot tell if H_0 is rejected because $\mu_1 \neq \mu_2$, or because $\mu_3 \neq \mu_5$, or because all five population means are unequal, and so on. (Techniques do exist for deciding among the various possibilities, but they are beyond the scope of this book. The interested reader is referred to Edwards.*)

THE GENERAL LOGIC OF ANOVA

Since it may seem paradoxical to test a null hypothesis about *means* by testing *variances*, the general logic of ANOVA will be discussed before proceeding to the computational procedures. The ANOVA procedure is based on a mathematical proof that the sample data can be made to yield two independent estimates of the population variance:

1. Within-group (or "error") variance estimate. This estimate is based on how different each of the scores in a given sample (or *group*) is from other scores in the same group.

2. Between-group variance estimate. This estimate is based on how different the *means* of the various samples (or *groups*) are from one another.

If the samples all come from the same normally distributed population (or from normally distributed populations with equal

* Edwards, A. L., *Experimental design in psychological research* (3rd Ed.), New York: Holt, 1969.

means and variances), it can be proved mathematically that the between-group variance estimate and the within-group variance estimate will be about equal to each other and to the population variance (σ^2). The larger the between-group variance estimate is in comparison to the within-group variance estimate, the more likely it is that the samples do *not* come from populations with equal means.

As an illustration, consider the hypothetical data in Table 14.1A. There are five groups (Group 1 = largest dose of caffeine, Group 5 = placebo) each containing five people, so there are 25 subjects in all, and the entries in the table represent the score of each person

Table 14.1   *Two possible outcomes of the caffeine experiment (hypothetical data)*

A. Within-group variation and between-group variation are about equal

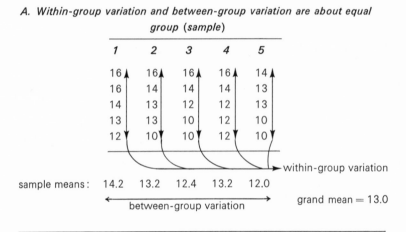

	group (sample)			
1	2	3	4	5
16	16	16	16	14
16	14	14	14	13
14	13	12	12	13
13	13	10	12	10
12	10	10	12	10

within-group variation

sample means: 14.2   13.2   12.4   13.2   12.0

between-group variation   grand mean = 13.0

B. Between-group variation much greater than within-group variation

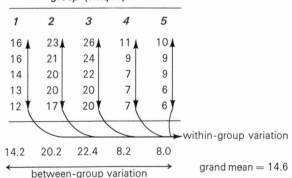

	group (sample)			
1	2	3	4	5
16	23	26	11	10
16	21	24	9	9
14	20	22	7	9
13	20	20	7	6
12	17	20	7	6

within-group variation

sample means: 14.2   20.2   22.4   8.2   8.0

between-group variation   grand mean = 14.6

on the English test. There is some between-group variation and some within-group variation. The means of the five groups may differ because caffeine does have an effect on test scores, or merely because they are affected by the presence of sampling error (that is, the accident of which cases happened to be in each sample), or for both reasons. Whether to retain or reject the null hypothesis that the samples all come from the same population is decided by applying ANOVA procedures to the data; inspection of Table 14.1A should suggest that the between-group variation and within-group variation are about equal, as would be expected if H_0 is true.

What if caffeine does have an effect? Suppose that one particular dosage causes an increase (or a decrease!) in test scores (that is, has about the same effect on all subjects receiving that dosage). The variability *within* that group will not be affected (remember, adding a constant to all scores, or subtracting a constant from all scores, does *not* change the variance or standard deviation), so neither will the within-group variance estimate. The mean of this group, however, will be higher than the mean of the other groups, so the *between*-group variance estimate will increase. Similarly, if several different dosages have different effects on test scores (for example, some dosages increase test scores more than others), the within-group variance estimate will not increase but the between-group variance estimate will be larger; the greater the differences among the group means, the larger will be the between-group variance estimate. As an illustration, consider the data in Table 14.1B. These data were obtained from the data in Table 14.1A as follows: scores in Group 1 were not changed; 7 was added to each score in Group 2; 10 was added to each score in Group 3; 5 was subtracted from each score in Group 4; and 4 was subtracted from each score in Group 5. The variance of each of the groups in Table 14.1B, and hence the within-group variance estimate, is equal to that of Table 14.1A, but the between-group variance estimate is much greater in the case of Table 14.1B. Thus, it is more likely that the five samples in Table 14.1B do *not* come from the same population.

To clarify further the meaning of between-group variation and within-group variation, consider the score of the first person in Group 1 of Table 14.1A, which is equal to 16. The total variation of his score from the grand mean of 13.0 (symbolized as $X - \bar{X}$, where $\bar{X}$ is the grand mean) is equal to $16 - 13$ or 3 raw score units. This total variation can be subdivided into two components, variation of the score from the mean of the first group (in symbols,

$X - \bar{X}_1$, where $\bar{X}_1$ is the mean of the first group) and variation of the mean of the first group from the total mean $(\bar{X}_1 - \bar{X})$:

$\bar{X}$	$\bar{X}_1$	X
13.0	14.2	16

$$\bar{X}_1 - \bar{X} = 1.2 \qquad X - \bar{X}_1 = 1.8$$

$$X - \bar{X} = (X - \bar{X}_1) + (\bar{X}_1 - \bar{X}) = 1.2 + 1.8 = 3.0$$

Thus, this person's score deviates by 1.8 points from the mean of Group 1. This is " error " in the sense that it cannot be explained by the caffeine variable, since all people in Group 1 received the same dosage. It simply reflects the basic variance of the English test scores. The difference between the mean of Group 1 and the grand mean, equal to 1.2 points, reflects how much the dosage given Group 1 caused the mean test score of this group to differ from the grand mean *plus* this same basic variance of English test scores. In sum, the three deviation scores involving this person are:

1. total deviation $= 3.0$
2. within-group (from own group) deviation (*error*) $= 1.8$
3. between-group (from own group to grand mean) deviation $= 1.2$.

The total variance, the within-group variance estimate, and the between-group variance estimate simply express the magnitude of these deviations, respectively, for everyone in the experiment. It can readily be shown that the total variability is equal to the sum of the between-group variance estimate and the within-group variance estimate.*

The between-group variance estimate includes *both* the effects of the caffeine (if any) *and* error variance; the group means will be affected both by the experimental treatment and by the basic English test or error variance of the scores on which they are based. The within-group variance estimate, however, reflects solely the error variance. The effect of the caffeine can therefore be determined by computing the following F ratio†:

$$F = \frac{\text{treatment variance} + \text{error variance}}{\text{error variance}}$$

$$= \frac{\text{between-group variance estimate}}{\text{within-group variance estimate}}$$

---

* The proof is given in the Appendix at the end of this chapter.
† If there are only two groups, the F test and the procedures given in Chapter 10 yield similar results. In fact, in this instance, $F = t^2$.

If H_0 is true, there is no treatment variance and the between-group and within-group variance estimates will be approximately equal, so that F will be approximately equal to 1.0. The more F is greater than 1.0, the more sure you can be that caffeine does have an effect on test scores, and you reject H_0 when F is so large as to have a probability of .05 or less of occurring if H_0 is true.

Values of F less than 1.0 would automatically indicate that H_0 should be retained, since the between-group variance estimate is smaller than the within-group variance estimate. Since the value of F expected if H_0 is true is 1.0, one might conceive of a *significantly small* value of F (for example, 0.2). There would be no obvious explanation for such a result other than chance, or perhaps failure of one of the assumptions underlying the F test. Thus, you would retain H_0 in such instances, and check the possibility that some systematic factor like nonrandom sampling has crept in.

COMPUTATIONAL PROCEDURES

SUMS OF SQUARES

The first step in the ANOVA design is to compute the *sum of squares between groups* (symbolized by SS_B), the *sum of squares within groups* (symbolized by SS_W), and the *total sum of squares* (symbolized by SS_T). A *sum of squares* is nothing more than a sum of squared deviations.

1. Total sum of squares (SS_T). The definition formula for the total sum of squares is

$$SS_T = \sum (X - \bar{X})^2$$

where

$\bar{X}$ = grand mean for all observations in the experiment

$\sum$ = summation across all *observations*

In Table 14.1A, for example,

$$SS_T = (16 - 13)^2 + (16 - 13)^2 + (14 - 13)^2 + \cdots + (13 + 13)^2$$
$$+ (10 - 13)^2 + (10 - 13)^2$$
$$= 3^2 + 3^2 + 1^2 + \cdots + 0^2 + (-3)^2 + (-3)^2$$
$$= 100$$

As was the case with the variance (Chapter 5), however, it is usually easier to compute SS_T by using a short-cut computing formula:

$$SS_T = \sum X^2 - \frac{(\sum X)^2}{N}$$

where N = total number of observations in the experiment.

For the data in Table 14.1A,

$$\sum X^2 = 16^2 + 16^2 + 14^2 + \cdots + 13^2 + 10^2 + 10^2$$
$$= 4325$$
$$\frac{(\sum X)^2}{N} = \frac{(325)^2}{25} = 4225$$
$$SS_T = 4325 - 4225$$
$$= 100$$

2. Sum of squares between groups (SS_B). The definition formula for the sum of squares between groups is

$$SS_B = \sum N_G (\bar{X}_G - \bar{X})^2$$

where

N_G = number of scores in group G
$\bar{X}_G$ = mean of group G
$\bar{X}$ = grand mean
$\sum$ = summation across all the *groups* (*not* the subjects)

As we have seen, the between-groups sum of squares deals with the difference between the mean of each group and the grand mean; and the squared difference is effectively counted only once for each person in the group (samples need not be of equal size). In Table 14.1A,

$$SS_B = 5(14.2 - 13.0)^2 + 5(13.2 - 13.0)^2 + 5(12.4 - 13.0)^2$$
$$\qquad + 5(13.2 - 13.0)^2 + 5(12.0 - 13.0)^2$$
$$= 5(1.2)^2 + 5(.2)^2 + 5(-.6)^2 + 5(.2)^2 + 5(-1.0)^2$$
$$= 14.4$$

A simpler computing formula for SS_B is

$$SS_B = \frac{(\sum X_1)^2}{N_1} + \frac{(\sum X_2)^2}{N_2} + \cdots + \frac{(\sum X_k)^2}{N_k} - \frac{(\sum X)^2}{N}$$

where

$\sum X_1$ = sum of scores in group 1, $\sum X_2$ = sum of scores in group 2, etc.
k = the number of groups (hence, the last group)
N_1 = number of scores in group 1, N_2 = number of scores in group 2, etc.

For example, in Table 14.1A,

$$SS_B = \frac{(71)^2}{5} + \frac{(66)^2}{5} + \frac{(62)^2}{5} + \frac{(66)^2}{5} + \frac{(60)^2}{5} - \frac{(325)^2}{25}$$
$$= 4239.4 - 4225$$
$$= 14.4$$

3. Sum of squares within groups (SS_W). The definition formula for the sum of squares within groups is

$$SS_W = \sum (X_1 - \bar{X}_1)^2 + \sum (X_2 - \bar{X}_2)^2 + \cdots + \sum (X_k - \bar{X}_k)^2$$

where

$X_1 =$ score in group 1, $X_2 =$ score in group 2, etc.
$k =$ last group
$\bar{X}_1 =$ mean of first group, $\bar{X}_2 =$ mean of second group, etc.
$\sum =$ summation across the N_G cases of the group in question

In Table 14.1A,

$$\begin{aligned}
SS_W = &\ (16 - 14.2)^2 + (16 - 14.2)^2 + (14 - 14.2)^2 + (13 - 14.2)^2 \\
&+ (12 - 14.2)^2 + (16 - 13.2)^2 + (14 - 13.2)^2 + (13 - 13.2)^2 \\
&+ (13 - 13.2)^2 + (10 - 13.2)^2 + (16 - 12.4)^2 + \cdots \\
&+ (10 - 12.0)^2 + (10 - 12.0)^2 \\
= &\ 85.6
\end{aligned}$$

That is, the difference between each score and the mean of the group containing that score is squared; once this has been done for every score, the results are summed. This is the most tedious of the three sums of squares to compute, and it is possible to make use of the fact that

$$SS_T = SS_B + SS_W$$

That is, the sum of squares between groups and the sum of squares within groups must add up to the total sum of squares. (See the Appendix at the end of this chapter.) Therefore, the within-group sum of squares can readily be computed as follows:

$$\begin{aligned}
SS_W &= SS_T - SS_B \\
&= 100 - 14.4 \\
&= 85.6
\end{aligned}$$

You should be careful in using this shortcut, however, because it provides no check on computational errors. If SS_W is found directly, the above formula can be used as a check on the accuracy in computation.

MEAN SQUARES

Next, SS_B and SS_W are each divided by the appropriate degrees of freedom. The values thus obtained are called *mean squares*, and are estimates of the population variance. The degrees of freedom between groups (symbolized by df_B) is equal to

$$df_B = k - 1 \quad \text{where} \quad k = \text{number of groups}$$

The degrees of freedom within groups (symbolized by df_W) is equal to

$$df_W = N - k \qquad \text{where} \quad N = \text{total number of observations}$$

This is equivalent to obtaining the degrees of freedom for each group separately ($N_G - 1$) and then adding the df across all groups. The *total* degrees of freedom, helpful as a check on the calculations of df, is equal to $N - 1$.

In Table 14.1A, $df_B = (5 - 1) = 4$; $df_W = (25 - 5) = 20$. The total $df = N - 1$ or 24, which is in fact equal to $4 + 20$.

The mean squares between groups (symbolized by MS_B) and the mean squares within groups (symbolized by MS_W) are equal to:

$$MS_B = \frac{SS_B}{df_B}$$

$$MS_W = \frac{SS_W}{df_W}$$

Thus, in Table 14.1A,

$$MS_B = \frac{14.4}{4}$$

$$= 3.60$$

$$MS_W = \frac{85.6}{20}$$

$$= 4.28$$

THE F RATIO

Having computed the mean squares, the last step is to compute the F ratio, where

$$F = \frac{MS_B}{MS_W}$$

In Table 14.1A,

$$F = \frac{3.60}{4.28}$$

$$= 0.84$$

TESTING THE F RATIO FOR STATISTICAL SIGNIFICANCE

Just as there are different t distributions for different degrees of freedom, so are there different F distributions for all combinations of different df_B and df_W. Although the specific shape of an F distribution depends on df_B and df_W, all F distributions are positively skewed. Two of them are illustrated in Figure 14.1.

In order to obtain the minimum value of F needed to reject H_0, you refer to Table F in the Appendix and consult the *column* corresponding to df_B (the degrees of freedom in the *numerator* of the F ratio) and the *row* corresponding to df_W (the degrees of freedom in the *denominator* of the F ratio). The critical F values for 4 df (numerator) and 20 df (denominator) are:

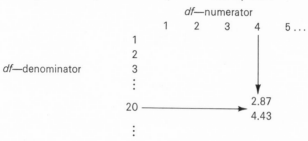

Figure 14.1   *F distributions for $df_B = 4$ and $df_W = 20$, and for $df_B = 6$ and $df_W = 6$.*

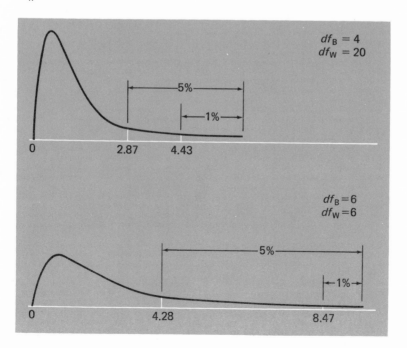

(The critical values for 20 df (numerator) and 4 df (denominator) are *not* the same; be sure to enter the table in the proper place.) The smaller of the two values is the .05 criterion; the larger, the .01 criterion. In the case of Table 14.1A, the obtained value of 0.84 is less than the critical value of 2.87; therefore, using $a = .05$, you retain H_0 and conclude there is not sufficient reason to believe that there are any differences among the five population means. (Since the computed value of F is less than 1.0, you could have reached this conclusion without bothering to consult the table in this instance; the *expected* (mean) value of any F distribution is close to 1, so no value of 1 or less can be significant.)

THE ANOVA TABLE

It is customary to summarize the results of an analysis of variance in a table such as the one shown in Table 14.2. Note that the

Table 14.2   *Summary of one-way ANOVA of caffeine experiment*

source of variation	SS	df	MS	F
between groups	14.4	4	3.60	0.84
within groups (error)	85.6	20	4.28	

between-groups source of variation is listed first, and that the value of F is entered at the extreme right of the between-groups row.

ONE-WAY ANOVA WITH UNEQUAL SAMPLE SIZES

An example of one-way ANOVA for samples of unequal size is shown in Table 14.3. The problem is whether or not students in three different eastern colleges in the United States differ in their attitudes toward student participation in determining college curricula. Three samples are randomly obtained from the three populations, and each person is given an attitude measure (the higher the score, the more positive the attitude). The statistical analysis is shown in Table 14.3; the computed value of F is 4.73. The critical value of F obtained from the F table for 2 and 24 degrees of freedom and $a = .05$ is 3.40. Since the obtained F value is larger than the critical value obtained from the table, you reject H_0 and conclude that it is unlikely that the three samples come from populations with equal means. (You may *not* draw formal conclusions about any two of the population means without additional statistical procedures; H_1 states only that the null hypothesis of $\mu_1 = \mu_2 = \mu_3$ is not true.)

Table 14.3 *ANOVA analysis of attitudes of students from three colleges in the Eastern U.S. to student participation in determining college curricula*

group			
1	*2*	*3*	
15	17	6	$H_0: \mu_1 = \mu_2 = \mu_3$
18	22	9	$H_1: H_0$ is untrue
12	5	12	$a = .05$
12	15	11	
9	12	11	
10	20	8	
12	14	13	
20	15	14	
	20	7	
	21		

$N_1 = 8 \qquad N_2 = 10 \qquad N_3 = 9 \qquad N = 27$

$\sum X_1 = 108 \qquad \sum X_2 = 161 \qquad \sum X_3 = 91 \qquad \sum X = 360 \qquad \sum X^2 = 5402$

$$SS_T = 5402 - \frac{(360)^2}{27}$$

$$= 5402 - 4800 = 602$$

$$SS_B = \frac{(108)^2}{8} + \frac{(161)^2}{10} + \frac{(91)^2}{9} - \frac{(360)^2}{27}$$

$$= 4970.21 - 4800 = 170.21$$

$$SS_W = 602 - 170.21 = 431.79$$

source of variation	SS	df	MS	F
between groups	170.21	$(3-1) = 2$	$170.21/2 = 85.10$	$85.10/17.99$ $= 4.73$
within groups (error)	431.79	$27 - 3 = 24$	$431.79/24 = 17.99$	

SOME COMMENTS ON THE USE OF ANOVA

UNDERLYING ASSUMPTIONS

The procedures discussed in this chapter assume that observations are *independent*; the value of any observation should *not* be related in any way to that of any other observation, as, for example, would be the case if the three samples were selected to be equal on average IQ. It is also formally assumed that the variances are equal for all treatment populations (that is, homogeneous), and that the populations are normally distributed. ANOVA is robust with regard to these assumptions, however, and will yield accurate results even if population variances are not homogeneous (provided that sample sizes are about equal) and even if population shapes depart moderately from normality.

MEASURES OF STRENGTH OF RELATIONSHIP

As was pointed out in previous chapters, statistical significance does not necessarily imply a strong relationship. It is therefore desirable to have a measure of the strength of the relationship between the independent and dependent variable in addition to the test of significance. One such measure is provided by ε (epsilon):

$$\varepsilon = \sqrt{\frac{df_B(F-1)}{df_B F + df_W + 1}}$$

For example, consider the data in Table 14.3, where $F = 4.73$, $df_B = 2$, and $df_W = 24$. Then, ε is equal to:

$$\varepsilon = \sqrt{\frac{(2)(4.73-1)}{(2)(4.73)+24+1}}$$

$$= \sqrt{\frac{7.460}{34.46}}$$

$$= \sqrt{.2165}$$

$$= .47$$

Epsilon bears the same relationship to F that r_{pb} bears to t (see Chapter 12). Thus, you would conclude that there is a fairly strong relationship in the sample between college and attitudes to student participation in determining college curricula. Its significance is, of course, given by the F.

OTHER FORMS OF ANOVA

There are many variations of ANOVA. As we have seen, the one-way design involves one independent variable, two or more groups, and the null hypothesis that the groups selected at random come from populations with equal means on the dependent variable. One of the great advantages of ANOVA is that it is possible to include *more than one independent variable* in the statistical analysis. Such techniques, called factorial designs, range from fairly straightforward to complex; an introduction to factorial design is presented in the following chapter.

SUMMARY

Analysis of variance (ANOVA) permits null hypotheses to be tested which involve the means of three or more samples (groups); one-way ANOVA deals with one independent variable. Total variance is partitioned into two sources: *between-group variance* and *within-group (error) variance*. These are compared by using the *F* ratio to determine whether or not the independent variable has an effect on the dependent variable. *Epsilon* is useful as a measure of the strength of the relationship between the independent and dependent variable.

APPENDIX TO CHAPTER 14: PROOF THAT TOTAL VARIANCE IS EQUAL TO THE SUM OF BETWEEN-GROUP AND WITHIN-GROUP VARIANCE

It is obvious that

$$X - \bar{X} = X - \bar{X} + \bar{X}_G - \bar{X}_G$$

Rearranging the terms,

$$X - \bar{X} = (X - \bar{X}_G) + (\bar{X}_G - \bar{X})$$

Squaring both sides of the equation and summing over *all* people gives

$$\sum (X - \bar{X})^2 = \sum (X - \bar{X}_G)^2 + \sum (\bar{X}_G - \bar{X})^2 + 2 \sum (X - \bar{X}_G)(\bar{X}_G - \bar{X})$$

For any one group, $(\bar{X}_G - \bar{X})$ is a constant, and the sum of deviations about the group mean must equal zero. The last term is therefore always equal to zero, leaving

$$\sum (X - \bar{X})^2 = \sum (X - \bar{X}_G)^2 + \sum (\bar{X}_G - \bar{X})^2$$

Dividing both sides of this equation by $N - 1$ would yield the definition formula for the variance:

$$s^2 = s_B^2 + s_W^2$$

For ANOVA purposes, however, the sums of squares are instead divided by the appropriate degrees of freedom.

15 introduction to factorial design: two-way analysis of variance

The one-way analysis of variance presented in Chapter 14 is used to investigate the relationship of a single independent variable to a dependent variable, where the independent variable has two or more levels (that is, groups). For example, the caffeine experiment in Chapter 14 dealt with the effect of five different dosages of caffeine (five levels of the independent variable) on performance on an English test (the dependent variable).

The *factorial design* is used to study the relationship of *two or more* independent variables (called *factors*) to a dependent variable, where each factor has two or more levels. For example, suppose you are interested in the relationship between four different dosages of caffeine (four levels: large, moderate, small, zero) and sex (two levels: male and female) to scores on a 20-item English test. There are several hypotheses of interest: different dosages of caffeine may affect test scores; males and females may differ in test performance; certain caffeine dosages more than others may affect test scores more for one sex than for the other. These hypotheses may be evaluated in a single statistical framework by using a factorial design; this example would be called a "two-way" analysis of variance to indicate two independent variables. This example could also be labeled as a 4×2 factorial design because there are four levels of the first independent variable and two levels of the second independent variable.

The logic of the factorial design follows from the logic of the more simple one-way design. The total variance is partitioned into within-group variance (error) estimate and between-group variance estimate. In the factorial design, however, the between-group variance estimate is itself partitioned into several parts: variation due to the first factor, variation due to the second factor, and variation due to the joint effects of the two factors (called the *interaction*). (See Figure 15.1.) An example of an interaction

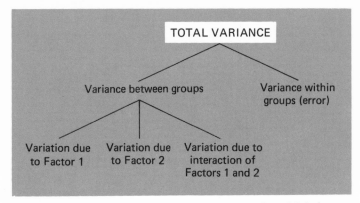

Figure 15.1   *Partitioning of variance in a two-way factorial design.*

effect would be if a particular dosage of caffeine improved test scores for males but *not* for females, while other dosages had no effect on test scores for either sex. (The interaction, a particularly important feature of factorial designs, will be discussed in detail later in this chapter.) It can be shown algebraically that the total sum of squares is equal to the sum of these various parts: (1) the within-group (error) sum of squares, (2) sum of squares due to factor 1, (3) sum of squares due to factor 2, and (4) sum of squares due to the interaction of factors 1 and 2. Thus, a factorial design makes it possible to break down the total variability into *several* meaningful parts. That is, it permits several possible explanations as to why people are different on the dependent variable; and, as was pointed out in Chapter 5, explaining variation—why people differ from one another—is the *raison d'etre* of the behavioral scientist.

COMPUTATIONAL PROCEDURES

The outline of the computational procedure for the two-way factorial design is as follows:

1. A.   Compute SS_T.
    B.   Compute SS_B.
    C.   Subtract SS_B from SS_T to obtain SS_W (error).
    D.   Compute SS_1 (the sum of squares for factor 1).
    E.   Compute SS_2 (the sum of squares for factor 2).
    F.   Subtract SS_1 and SS_2 from SS_B to obtain the sum of squares for the interaction of factors 1 and 2 ($SS_{1 \times 2}$).
2. Convert the sums of squares in C, D, E, and F above to mean squares by dividing each one by the appropriate degrees of freedom.

3. A. Test the mean square for factor 1 (MS_1) for statistical significance by computing the appropriate F ratio.

B. Test the mean square for factor 2 (MS_2) for statistical significance by computing the appropriate F ratio.

C. Test the mean square for interaction ($MS_{1 \times 2}$) for statistical significance by computing the appropriate F ratio.

Thus, the two-way factorial design permits you to test three null hypotheses—one concerning the effect of factor 1, one concerning the effect of factor 2, and one concerning the *joint effect* of factor 1 and factor 2—in a single statistical framework.

As an illustration, consider the hypothetical results for the caffeine experiment shown in Table 15.1. There are five observations in each cell; for example, the scores of 6, 15, 12, 12, and 13 are the test scores of five men who received a large dose of caffeine. (Techniques for analyzing factorial designs with unequal numbers of scores in different cells are beyond the scope of this book. Also, the procedures in this chapter require more than one observation in each cell.) Just as in the one-way design, the within-group variance estimate is based on the variability *within* each of the eight cells. Variation due to the caffeine factor is reflected by the variability across the four *column* means, while variation due to the sex factor is reflected by the variability (that is, difference) of the two *row* means.

SUMS OF SQUARES

1. Total sum of squares (SS_T). The total sum of squares is computed in the same way as in Chapter 14:

$$SS_T = \sum X^2 - \frac{(\sum X)^2}{N}$$

where

N = total number of observations

$\sum$ = summation across all *observations*

In Table 15.1,

$$\sum X^2 = 6^2 + 15^2 + 12^2 + \cdots + 6^2 + 9^2 + 9^2$$

$$= 4394$$

$$\frac{(\sum X)^2}{N} = \frac{(402)^2}{40} = 4040.1$$

$$SS_T = 4394 - 4040.1$$

$$= 353.9$$

Table 15.1   *Scores on a 20-item English test as a function of caffeine dosage and sex (4 × 2 factorial design)*

		caffeine dosage (factor 1)				row sums	row means
		large	moderate	small	zero		
sex (factor 2)	male	6	12	10	9	217	10.85
		15	10	13	10		
		12 (sum = 58)	12 (sum = 54)	15 (sum = 60)	7 (sum = 45)		
		12	13	12	12		
		13	7	10	7		
	female	10	9	12	4	185	9.25
		13	7	13	7		
		4 (sum = 41)	10 (sum = 46)	15 (sum = 63)	6 (sum = 35)		
		9	7	10	9		
		5	13	13	9		
column sums		99	100	123	80	grand sum = 402	
column means		9.9	10.0	12.3	8.0	grand mean = 10.05	

2. Sum of squares between groups (SS_B). We can ignore for a moment the fact that this is a factorial design, and simply treat the data in Table 15.1 as 8 groups and find SS_B as in Chapter 14:

$$SS_B = \frac{(\sum X_1)^2}{N_1} + \frac{(\sum X_2)^2}{N_2} + \cdots + \frac{(\sum X_k)^2}{N_k} - \frac{(\sum X)^2}{N}$$

For the data in Table 15.1, $k = 8$, and

$$SS_B = \frac{(58)^2}{5} + \frac{(54)^2}{5} + \cdots + \frac{(63)^2}{5} + \frac{(35)^2}{5} - \frac{(402)^2}{40}$$

$$= \frac{20,896}{5} - \frac{161,604}{40}$$

$$= 4179.2 - 4040.1$$

$$= 139.1$$

3. Sum of squares within groups (error) (SS_W). The within-groups sum of squares may be found by subtraction:

$$SS_W = SS_T - SS_B$$
$$= 353.9 - 139.1$$
$$= 214.8$$

4. Sum of squares for factor 1 (SS_1). Let us define caffeine as factor 1. The sum of squares for the caffeine factor, *which ignores sex differences*, is *

$$SS_1 = \sum \frac{(\text{sum of each } \textit{column})^2}{(N \text{ in each column})} - \frac{(\sum X)^2}{N}$$

where

$$\sum = \text{summation across all } \textit{columns}.$$

That is, if caffeine has an effect on test scores (ignoring sex), the means (and hence the sums) of the *columns* of Table 15.1 should show high variability. The value of $(\sum X)^2/N$ has already been found to be equal to 4,040.1. Then,

$$SS_1 = \frac{(99)^2}{10} + \frac{(100)^2}{10} + \frac{(123)^2}{10} + \frac{(80)^2}{10} - 4040.1$$

$$= 4133 - 4040.1$$

$$= 92.9$$

* The formulas for many of the terms in factorial designs involve procedures of double and triple summation. Rather than introduce this notation solely for this chapter, however, we will use a more verbal presentation.

5. Sum of squares for factor 2 (SS_2). Let us define sex as factor 2. The sum of squares for the sex factor, *which ignores differences in caffeine dosage*, is

$$SS_2 = \sum \frac{(\text{sum of each } row)^2}{(N \text{ in each row})} - \frac{(\sum X)^2}{N}$$

where

$\sum$ = summation across all *rows*.

That is, if sex has an effect on test scores (ignoring caffeine), the means (and hence the sums) of the *rows* of Table 15.1 should show high variability.

$$SS_2 = \frac{(217)^2}{20} + \frac{(185)^2}{20} - 4040.1$$
$$= 4065.7 - 4040.1$$
$$= 25.6$$

6. Sum of squares for interaction ($SS_{1\times2}$). The interaction sum of squares is part of the variability of the 8 cells and is obtained by subtraction:

$$SS_{1\times2} = SS_B - SS_1 - SS_2$$

In Table 15.1,

$$SS_{1\times2} = 139.1 - 92.9 - 25.6$$
$$= 20.6$$

MEAN SQUARES

The next step is to convert each sum of squares to an estimate of the population variance, that is, to an average or mean square by dividing by the appropriate degrees of freedom, as shown following and illustrated in Figure 15.2.

Figure 15.2 *Partitioning of degrees of freedom in caffeine experiment.*

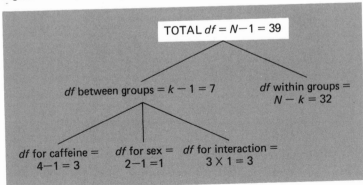

source	degrees of freedom	computation for Table 15.1
total	$N-1$	$df_T = 40 - 1 = 39$
within groups	$N-k$	$df_W = 40 - 8 = 32$
between groups	$k-1$	$df_B = 8 - 1 = 7$
Factor 1	(number of levels of factor 1) -1	$df_1 = 4 - 1 = 3$
Factor 2	(number of levels of factor 2) -1	$df_2 = 2 - 1 = 1$
interaction	$df_1 \times df_2$	$df_{1 \times 2} = 3 \times 1 = 3$

Note: N = total number of observations
k = number of cells

The first three values in the above table are the same as in the case of the one-way design (Chapter 14). The total degrees of freedom equals $N-1$, or one less than the total number of observations. The within-group degrees of freedom is equal to $N-k$, where k equals the number of cells (or groups); this is equivalent to obtaining the degrees of freedom for each cell (one less than the number of observations in the cell, or 4) and summing over all cells ($4 \times 8 = 32$). The between-group degrees of freedom equals one less than the number of cells. It is also necessary, however, to partition the between-group degrees of freedom in the same way as the between-group variance is partitioned. The degrees of freedom for factor 1 is one less than the number of levels of factor 1; in Table 15.1, there are four dosages of caffeine, so the degrees of freedom are $4-1 = 3$. Similarly, the degrees of freedom for factor 2, sex, are $2-1 = 1$. The degrees of freedom for the interaction of caffeine and sex is found by *multiplying* the degrees of freedom for each factor ($3 \times 1 = 3$).

The mean squares are then found by dividing each sum of squares by the corresponding degrees of freedom:

Mean square within groups:

$$(MS_W) = \frac{SS_W}{df_W} = \frac{214.8}{32} = 6.71$$

Mean square for caffeine:

$$(MS_1) = \frac{SS_1}{df_1} = \frac{92.9}{3} = 30.97$$

Mean square for sex:

$$(MS_2) = \frac{SS_2}{df_2} = \frac{25.6}{1} = 25.6$$

Mean square for interaction:

$$(MS_{1 \times 2}) = \frac{SS_{1 \times 2}}{df_{1 \times 2}} = \frac{20.6}{3} = 6.87$$

F RATIOS AND TESTS OF SIGNIFICANCE

The first null hypothesis to be tested is that the four caffeine groups come from populations with equal means. The mean square for the caffeine factor is divided by the mean square within groups to yield the following F ratio:

$$F = \frac{MS_1}{MS_w} = \frac{30.97}{6.71} = 4.62$$

The critical value from the F table for three degrees of freedom in the numerator and 32 degrees of freedom in the denominator and $a = .05$ is 3.30. Since the computed F of 4.62 is greater than this value, you reject H_0 and conclude that the four groups do *not* come from populations with equal means—that is, different dosages of caffeine *do* have an effect on the English test scores.

The second null hypothesis to be tested is that the two sexes come from populations with equal means. The mean square for the sex factor is divided by the mean square within groups to yield the following F ratio:

$$F = \frac{MS_2}{MS_w} = \frac{25.60}{6.71} = 3.82$$

The critical value from the F table for one degree of freedom in the numerator and 32 degrees of freedom in the denominator and $a = .05$ is 4.15. (Note that the critical values for the various F tests differ if the *df* differ, since they are different F distributions.) Since the computed value of 3.82 is less than the critical value, you retain H_0 and conclude that there is not sufficient reason to believe that different sexes perform differently on the English test.

The third null hypothesis to be tested is that the interaction effect is zero:

$$F = \frac{MS_{1 \times 2}}{MS_w} = \frac{6.87}{6.71} = 1.02$$

This computed value of F is less than the critical value obtained from the F table of 3.30 for 3 and 32 df and $\alpha = .05$, so you retain H_0 and conclude that there is not sufficient reason to reject the null hypothesis of no interaction effect.

ANOVA TABLE

The results of the above analysis of variance are summarized in Table 15.2. Note that the factors are identified by name for the

Table 15.2   *Summary of two-way ANOVA of caffeine experiment*

source	SS	df	MS	F
caffeine	92.90	3	30.97	4.62
sex	25.60	1	25.60	3.82
caffeine × sex	20.60	3	6.87	1.02
error	214.80	32	6.71	

convenience of the reader. Also, as was the case with the one-way analysis of variance (Chapter 14), within-group variation (error) is listed last, and no F value is listed for error because error is used as the denominator of the various F ratios and is not itself the subject of a statistical test.

THE MEANING OF INTERACTION

Interaction refers to the *joint* effect of two or more factors on the dependent variable. In the caffeine experiment, for example, the interaction effect of caffeine and sex refers to the effect of particular joint combinations of the two factors, such as male— high dosage, female—moderate dosage, and so forth and *not* the sum of the separate effects of the two factors. It is the joint effect *over and above* the sum of the separate effects.

A numerical illustration of interaction is given in Table 15.3, which summarizes the cell means for the caffeine experiment. The overall mean difference between the sexes, ignoring the caffeine factor, shows that the men averaged 1.6 points higher than the women. If the mean differences between males and females for each dose of caffeine were all equal to 1.6, the interaction effect would be zero; that is, sex differences in English test performance

Table 15.3   *Cell means for the data in Table 15.1*

		caffeine dosage				
		large	*moderate*	*small*	*zero*	*overall*
sex	*male*	11.6	10.8	12.0	9.0	10.85
	female	8.2	9.2	12.6	7.0	9.25
$\bar{X}_{male} - \bar{X}_{female}$		3.4	1.6	−0.6	2.0	1.60

Table 15.4   *Hypothetical cell means illustrating zero and large interaction effects*

A. Zero interaction

		caffeine dosage				
		large	*moderate*	*small*	*zero*	*overall*
sex	*male*	11.6	10.8	12.0	9.0	10.85
	female	10.0	9.2	10.4	7.4	9.25
$\bar{X}_{male} - \bar{X}_{female}$		1.6	1.6	1.6	1.6	1.60

B. Large interaction

		caffeine dosage				
		large	*moderate*	*small*	*zero*	*overall*
sex	*male*	11.6	10.8	12.0	9.0	10.85
	female	8.2	9.2	7.0	12.6	9.25
$\bar{X}_{male} - \bar{X}_{female}$		3.4	1.6	5.0	−3.6	1.60

would be exactly the same regardless of the dosage of caffeine. Also, the difference between mean test scores for any two dosages of caffeine would be the same for both sexes. (See Table 15.4A.) If the mean differences for each dose of caffeine were considerably different from one another, there would be a significant interaction effect; the caffeine factor would affect test scores differently for different sexes. (One possible illustration is shown in Table 15.4B.) Since the interaction effect for the data in Table 15.3 was found to be *not* significant, the differences among the observed mean differences of 3.4, 1.6, −0.6, and 2.0 are likely to have occurred by chance (random sampling error), so there is *not* sufficient reason to believe that there is an interaction effect in the population from which the samples in this experiment were drawn.

A graphic illustration of the examples in Table 15.4 is shown in Figure 15.3. When the interaction effect is zero (Figure 15.3A), the line connecting the points corresponding to the cell means for men follows the same pattern as the corresponding line for women. That is, the distance between the two lines is the same at all points. When there is an interaction effect (Figure 15.3B), sex differences are different from one point to another; in this example, females obtained higher scores under the placebo (no caffeine) condition, while men obtained higher scores under the other three conditions—particularly the small caffeine dosage.

Two different kinds of interaction are shown for a 2×2 factorial design in Figure 15.4. In Figure 15.4A, men obtain higher average test scores than women when caffeine is administered, while women obtain higher average scores than men when the placebo is given. In Figure 15.4B, caffeine has little if any effect for women (the cell means for the caffeine and placebo conditions are about equal), but men perform better on the test when given caffeine. That is, sex differences are greater under the caffeine

Figure 15.3 *Graphic representation of data in Table 15.4.*

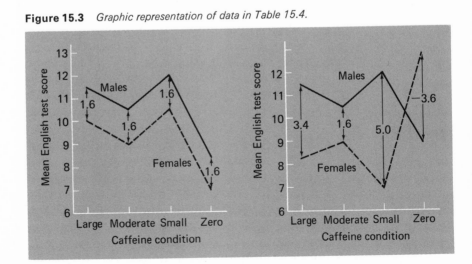

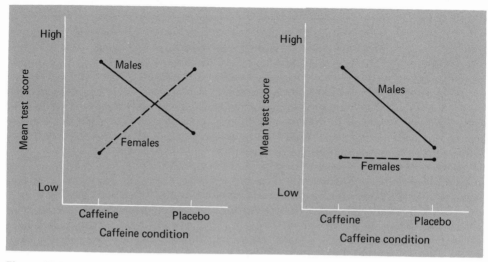

Figure 15.4   *Two kinds of interaction effect for caffeine and sex on test scores (2 × 2 factorial design).*

condition than under the placebo condition. Thus, there is a "reversal" effect in Figure 15.4A; whereas in Figure 15.4B, caffeine affects test scores only for one sex.

It is important to keep in mind that just as the effects of sex and dosage level are independent of each other in this design, so is the interaction independent of each of the factors. What this mutual independence of these three kinds of effects means is that any combination of them can prove to be significant. For example, the interaction can prove to be significant when neither, or either, or both of the factors are significant. Thus, the *average* effect of dosage level (combining sexes) and the *average* effect of sex (combining dosage levels) might be zero, while males might be much higher than females at two dosage levels and much lower at two others; this would result in neither factor significant, but interaction significant. Or, dosage level might result in different mean scores (combining sexes), while sexes (combining dosage levels) are the same *on the average*. The sex *difference* may still markedly favor males at one dosage level and slightly favor females at the other three, resulting in a significant interaction.

Much more could be said about interaction, and about factorial design. There are many other forms of factorial design; for example, procedures exist for including three or more independent variables in a single design, or for including several scores obtained from the same people. A discussion of such techniques, however, is best reserved for advanced texts. Suffice it to say that the factorial design in its various forms, including as it does the important concept of interaction, is one of the most valuable statistical tools at the disposal of the behavioral scientist.

SUMMARY

In two-way ANOVA (or two-way *factorial design*), one of many complex forms of ANOVA, there are *two* independent variables. The statistical analysis makes possible a significance test (using the *F* ratio) of the effect of *each* independent variable, and of the effect of the *interaction* of the two variables—that is, the *joint* effect of the two variables over and above the separate effects of each one.

16 chi square

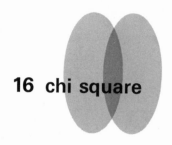

CHI SQUARE AND GOODNESS OF FIT: ONE-VARIABLE PROBLEMS

Consider the following problems:

1. A manufacturer produces three new shades of lipstick—cherry, pink, and apple blossom—and he wonders whether or not they will be equally popular among female consumers. To test the hypothesis of equal preference, he obtains a random sample of 177 women and asks them which *one* of the three colors they like best. He finds that 65 women prefer cherry, 60 prefer pink and 52 prefer apple blossom. Are these results sufficient reason to reject the null hypothesis that equal numbers of women *in the population* prefer each of the three colors? Or, are these results likely to occur as a result of sampling error if the preferences in the population are in fact equal, in which case the null hypothesis of equality in the population should be retained?

2. A college professor believes that most students in his college would like to eliminate final examinations. He obtains a random sample of students and finds that 160 would like to do away with final exams, while 115 would not. Can he reject the null hypothesis that the issue divides the student population equally?

3. A new proposal favored by students at a particular college requires a two-thirds vote of the faculty for approval. A random sample of 100 faculty members is obtained, and 55 favor the proposal while 45 are opposed. What conclusion should be reached with regard to the null hypothesis that two-thirds of the faculty favor the proposal?

In problems in previous chapters, using statistics such as *t* and *F*, groups were compared in terms of the *amount* of a given characteristic. For example, in Chapter 10, the experimental group given caffeine had greater test scores on the average ($\bar{X} = 81.0$) than the control group not given caffeine ($\bar{X} = 78.0$), and the *t* test dealt with the average amount of points scored by each group. It not infrequently happens, however, that the data consist only of the *number of objects* (for example, persons) *falling in any one of a number of categories*, and information about the amount of the given characteristic is not relevant or not available. That is, the data are in terms of *frequencies* (number of objects), and the empirically observed frequencies are compared to frequencies expected on the basis of some hypothesis. For example, the data in the three problems given at the beginning of this chapter consist of frequencies or head counts—that is, *how many* prefer a particular alternative—and the objective is to decide whether or not it is reasonable to conclude that the population frequencies are distributed in a specified way. Thus, in problem 1, the manufacturer knows how many women in the sample prefer one shade of lipstick to the other two, and he would like to know if it is reasonable to conclude that the population from which his sample was randomly drawn is equally divided with regard to preference for the three colors.

The testing of hypotheses about frequency data falling in category-sets is similar in strategy to testing differences between an observed sample mean and the hypothesized value of the population mean. The question asked by the investigator dealing with frequencies is: Can the differences between the observed frequencies (symbolized as f_o) and the frequencies expected *if* the null hypothesis is true (symbolized as f_e) be attributed to random sampling fluctuations, or are the differences likely to be due to nonchance factors—that is, to the falsity of the null hypothesis? The statistic used to test such hypotheses is *chi square* (χ^2), defined by

$$\chi^2 = \sum \frac{(f_o - f_e)^2}{f_e}$$

where f_o = observed frequency
   f_e = expected (null-hypothetical) frequency
   $\sum$ is taken over all the categories

If the differences between the observed frequencies and the expected frequencies are small, χ^2 will be small. The greater the difference between the observed frequencies and those expected under the null hypothesis, the larger χ^2 will be. If the differences between observed and expected values are so large collectively as to occur by chance only .05 or less of the time when the null hypothesis is true, the null hypothesis is rejected.

The three problems given at the beginning of this chapter are all concerned with differences in choice among different categories (or levels) of a *single* variable. Thus, in the first problem, the variable is lipstick preference, and each respondent selects one of three possible choices. In the second case, the variable is attitude toward final examinations, and the student makes one of two possible choices. In problem three, the variable under consideration is attitude toward the new proposal, for which there are also two alternatives.

In all three problems, the expected frequencies must be stated before χ^2 can be determined. Considering problem 1 first, you first set up a table with the three alternatives and tabulate the observed frequencies. (See Table 16.1.) The null hypothesis

Table 16.1 *Chi-square test for lipstick choices*

color	observed frequency (f_o)	expected frequency (f_e)	$f_o - f_e$	$(f_o - f_e)^2$	$\dfrac{(f_o - f_e)^2}{f_e}$
cherry	65	59	6	36	.610
pink	60	59	1	1	.017
apple blossom	52	59	−7	49	.831

$$\chi^2 = \sum \frac{(f_o - f_e)^2}{f_e} = 1.458$$

states that the three colors are equally preferred, so *if* it is true you would expect the sample of 177 women to be equally divided among the three categories. Thus, the expected frequencies under the null hypothesis are 177/3 or 59, 59, and 59. (Note that the sum of the expected frequencies must be equal to the sum of the observed frequencies.) Is it likely or unlikely that observed frequencies of 65, 60, and 52 would occur if the population frequencies are exactly equal? In other words, are the observed frequencies of 65, 60, and 52 significantly different from the expected frequencies of 59, 59, and 59? To answer this question, χ^2 has been computed in Table 16.1. Note that *each* value of $(f_o - f_e)^2$ is divided by its own expected frequency (all the expected frequencies happen to be equal in this particular problem, but this will not always be true since it depends on the particular null hypothesis), and the resulting three values are then summed to obtain χ^2, which is equal to 1.46.

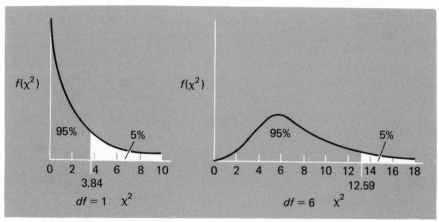

Figure 16.1 *Chi-square distribution for df = 1 and df = 6.*

In order to test the significance of χ^2 using a specified criterion of significance, the obtained value is referred to Table G in the Appendix with the appropriate degrees of freedom. Note that in this table, there is a different value of χ^2 for every *df*. Chi square, like *t*, yields a family of curves, with the shape of a particular curve depending on the *df*. (See Figure 16.1.) For the χ^2 distributions, however, the *df* are based on the number of *categories*, rather than on the sample size as in the case of the *t* distributions. In the one-variable case, the *df* are equal to $k - 1$, where k is equal to the number of categories of the variable. Thus, in the present problem, $df = 3 - 1 = 2$. For χ^2 to be significant at the .05 level, therefore, the obtained value must be equal to or greater than 5.99. Since the obtained value is only 1.46, the null hypothesis is retained; there is not sufficient reason to reject the null hypothesis that the frequencies in the population are equal. Therefore, we have no basis for concluding that any particular shade or shades is preferred.

The computation of chi square for the second problem posed at the beginning of this chapter is shown in Table 16.2. The null hypothesis specifies that the population of students are equally divided on the issue; that is, that 275/2 or 137.5 students will be in favor of final examinations and 137.5 students will be against final examinations. (Expected frequencies, since they are theoretical, are often in terms of fractions of a person.) The value of χ^2 is equal to 7.36. Since it is larger than the critical value of 3.84 obtained from Table G for $2 - 1 = 1$ *df* using the .05 criterion of significance, the null hypothesis is rejected—it is *not* reasonable to assume that the student population is evenly divided on this issue. By inspecting the f_o, it may be concluded that a majority are opposed to final examinations.

Table 16.2   *Chi-square test for attitude towards final examinations*

attitude	f_o	f_e	$f_o - f_e$	$(f_o - f_e)^2$	$\dfrac{(f_o - f_e)^2}{f_e}$
for	115	137.5	−22.5	506.25	3.682
against	160	137.5	22.5	506.25	3.682

$$\chi^2 = \sum \frac{(f_o - f_e)^2}{f_e} = 7.364$$

The analysis of the third problem is shown in Table 16.3. The null hypothesis specifies two-thirds of the faculty in favor of the proposition, so the expected frequency in favor is equal to $2/3 \times 100$ or 66.67. The expected frequency opposed to the proposition is $1/3 \times 100$ or 33.33. The obtained value of χ^2 of 6.13 is statistically significant (it is greater than the tabled value of 3.84 for $df = 1$); therefore, H_0 is rejected and the conclusion is that it is *not* reasonable to assume that two-thirds of the faculty population is in favor of the proposal. Observing the frequencies, we can conclude that *fewer* than two-thirds are in favor, and that therefore the proposal will not pass. (Lest unwarranted implications be drawn from Tables 16.2 and 16.3, we hasten to add that these data are entirely fictitious and are intended solely to illustrate the use of χ^2.)

Table 16.3   *Chi-square test for faculty attitudes to new student proposal*

attitude	f_o	f_e	$f_o - f_e$	$(f_o - f_e)^2$	$\dfrac{(f_o - f_e)^2}{f_e}$
for	55	66.67	−11.67	136.19	2.043
against	45	33.33	11.67	136.19	4.086

$$\chi^2 = \sum \frac{(f_o - f_e)^2}{f_e} = 6.129$$

SOME PRECAUTIONS INVOLVING THE USE OF χ^2

The χ^2 tests described above can only be performed when the observations are *independent*. That is, no response should be related to or dependent upon any other response. For example, it would be incorrect to apply χ^2 with $N = 100$ to the true–false responses of five schizophrenic patients to 20 questionnaire items because the 100 responses are not independent of each other; the 20 responses given by each must be assumed to be mutually related.

Second, any subject must fall in *one* and only one category. Thus, in the problem involving the preference for lipstick colors, each subject was asked to choose the one color she most preferred.

Third, the computations must be based on all the subjects in the sample. In problem 2, for example, a category of "for" as well as "against" must be included so that χ^2 is based on the total frequency of 275, the total size of the sample. As a check, the sum of the observed frequencies *must* be equal to the sum of the expected frequencies.

One final precaution is concerned with the size of the expected frequencies. Chi square is actually an approximate test for obtaining the probability values for the observed frequencies (that is, the probability of getting the observed frequencies if the null hypothesis is true.) This test is based on the assumption that within *any category*, sample frequencies are normally distributed about the population or expected value. Since frequencies cannot be negative, the assumption of normality is violated when expected population values are close to zero, since the sample frequencies cannot be much below the expected frequency while they can be much above it—an asymmetric distribution. Thus, this assumption is not troublesome when the expected frequencies are fairly large, but the smaller the expected frequencies, the less valid are the results of the χ^2 test. In fact, under certain conditions, you should *not* compute χ^2. For 1 *df*, *expected* frequencies should all be at least 5; otherwise χ^2 should not be used. For 2 *df*, expected frequencies should all exceed 2. With 3 or more *df*, if all expected frequencies but one are greater than or equal to 5 and if the one that does not is at least equal to 1, χ^2 is still a good approximation. In other words, the greater the *df*, the more lenient the criteria for expected frequencies.

CHI SQUARE AS A TEST OF INDEPENDENCE: TWO-VARIABLE PROBLEMS

In the preceding section, χ^2 was used to test some a priori hypothesis about expected frequencies—that is, a hypothesis about f_e values formulated prior to the experiment—which involved frequency data concerning a single variable. Chi square can also be used, however, *to test the significance of the relationship between two variables when data are expressed in terms of frequencies of joint occurrence.* For example, suppose you want to find out if men and women differ in their preference for the two major political parties, Democratic and Republican. If the two variables of sex and political party are *not* related (are *independent*), you would expect the proportion of men who prefer Democrats to be the same as the proportion of women who prefer Democrats. (See Table 16.4.) No a priori expected frequencies are involved; instead, you are dealing with the relationship between two variables, each of which may have any number of categories or levels (in this example, each variable has two levels), and the null hypothesis is that the two variables are independent, which (as you will see) implies a set of expected frequencies.

As an illustration, suppose you have a random sample of 111 registered male voters and a random sample of 69 registered female voters. You ask each individual to state which of the two major political parties is more preferred. With two categories of sex and two of political parties, there are four possible combined

Table 16.4 *Hypothetical illustration of perfectly independent relationship between sex and political preference, using frequency data*

		sex		
		female	*male*	total people
political preference	*Democrats*	46	74	120
	Republicans	23	37	60
	Total	69 +	111 =	180

66.67% of the women (46/69) prefer Democrats; 33.33% (23/69) prefer Republicans

66.67% of the men (74/111) prefer Democrats; 33.33% (37/111) prefer Republicans

38.33% of the people who prefer Democrats are women (46/120); 61.67% (74/120) are men

38.33% of the people who prefer Republicans are women (23/60); 61.67% (37/60) are men

Thus, there is no relationship between sex and political preference.

categories or *cells*: male–Democratic, male–Republican, female–Democratic, female–Republican. The results are shown in Table 16.5.

Table 16.5 *Table of frequencies relating sex to political preference (hypothetical data)*[a]

		sex		
		female	*male*	total people
political preference	Democrats	50 (46)	70 (74)	120
	Republicans	19 (23)	41 (37)	60
	Total	69 +	111 =	180

calculations

f_o	f_e	$f_o - f_e$	$(f_o - f_e)^2$	$\dfrac{(f_o - f_e)^2}{f_e}$
50	46	4	16	.348
70	74	−4	16	.216
19	23	−4	16	.696
41	37	4	16	.432
				$\chi^2 = 1.692$

[a] Within each cell, f_o is in the upper left-hand corner and f_e is in parentheses.

Thus, 50 women preferred the Democrats, 70 men preferred the Democrats, 19 women preferred the Republicans, and 41 men preferred the Republicans. The total for the first column (called the *marginal frequency* for that column) indicates that the sample contains 69 (that is, 50 + 19) women, while the marginal frequency for the second column shows that the sample contains 111 men. Similarly, the row marginal frequencies indicate that 120 persons of both sexes preferred the Democrats while 60 preferred the Republicans. The total sample size N is equal to 180.

The null hypothesis is stated in terms of the independence of the two variables, sex and political preference. Thus:

H_0: sex and political party are independent (*not* related)
H_1: sex and political party *are* related

The expected frequencies are those of Table 16.4—the frequencies predicated on the independence of the two variables. The procedure for computing the expected frequencies can be summarized as follows: for any cell, the expected value is equal to the product of the two marginal frequencies common to the cell (the row total times the column total) divided by the total N. That is,

$$f_e = \frac{(\text{row total})(\text{column total})}{N}$$

For example, in Table 16.5, the expected frequency for the female–Democratic cell is equal to

$$\frac{(120)(69)}{180} = 46$$

The computation of χ^2 is shown in Table 16.5. The resulting value of 1.69 is tested for statistical significance by referring to Table G in the Appendix with the proper df. For a two-variable problem, the df are equal to

$$df = (r-1)(c-1)$$

where

r = number of rows
c = number of columns

Thus, for the 2 × 2 table, the df are equal to $(2-1)(2-1) = 1$. The reason why there is one degree of freedom in a 2 × 2 table is as follows: Consider the expected frequency of 46 for the female–Democratic cell. Having computed this value, the expected frequency for the male–Democratic cell is fixed, that is, not free to vary. This is because the total of the expected frequencies in the first row of the table, like the total of the observed frequencies, must add up to 120—the marginal frequency for that row. Thus, the expected frequency for the male–Democratic cell is equal to

$$120 - 46 = 74$$

Similarly, the expected values in the first column must add up to the marginal frequency of 69. Thus, having computed the expected value of 46 for the female–Democratic cell, the expected frequency for the female–Republican cell is fixed; it must be

$$69 - 46 = 23$$

Finally, the expected frequency for the male–Republican cell is equal to

$$111 - 74 = 37$$

Thus, once any one f_e is known, all of the others are automatically determined. In other words, only one f_e is free to vary; the values of the others depend upon its value. Therefore, the 2×2 table has one degree of freedom.

In a larger table, all but one of the values of f_e in a given row or column are free to vary; once they are specified, the last one is fixed by virtue of the fact that the expected frequencies must add up to the marginal frequency. An illustration of the degrees of freedom for a 4×3 table is shown in Figure 16.2.

Returning to the problem in Table 16.5, the minimum value of χ^2 required to reject H_0 for $df = 1$ and $a = .05$ is 3.84. Since the obtained value of 1.69 is less than the tabled value, you would retain the null hypothesis and conclude that there is not sufficient reason to believe that the variables of sex and political preference are related.

The procedures discussed in this section can be used for two-variable problems with any number of levels for each variable. An example of a 3×4 problem is shown in Table 16.6. The expected frequency for a given cell is obtained by multiplying the row total for that cell by the column total and dividing by the total N. For example, f_e for the Black–Democratic cell is

$$\frac{(60)(138)}{300} = 27.60.$$

Since the computed value of χ^2 of 27.56 exceeds the tabled value for $(3 - 1)(4 - 1)$ or 6 df of 12.59, you reject H_0 and conclude that ethnic group and political preference *are* related.

COMPUTING FORMULA FOR 2×2 TABLES

For a 2×2 table, the following formula for χ^2 is equivalent to the one given previously and requires somewhat less work computationally:

Observed frequencies:

A	B
C	D

$$\chi^2 = \frac{N(AD - BC)^2}{(A+B)(C+D)(A+C)(B+D)}$$

Figure 16.2 *Illustration of degrees of freedom for a 4×3 table.*

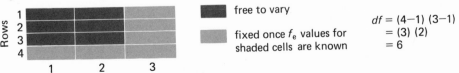

Table 16.6   *Table of frequencies relating ethnic group to political preference* (*hypothetical data*)

ethnic group	political preference				row total
	democratic	republican	liberal	conservative	
Black	36	8	14	2	60
White	84	72	18	26	200
Spanish-speaking	18	8	10	4	40
column total	138	88	42	32	300

calculations

f_o	f_e	$f_o - f_e$	$(f_o - f_e)^2$	$\dfrac{(f_o - f_e)^2}{f_e}$
36	27.60	8.40	70.56	2.56
8	17.60	−9.60	92.16	5.24
14	8.40	5.60	31.36	3.73
2	6.40	−4.40	19.36	3.03
84	92.00	−8.00	64.00	.70
72	58.67	13.33	177.69	3.03
18	28.00	−10.00	100.00	3.57
26	21.33	4.67	21.81	1.02
18	18.40	−0.40	.16	0.01
8	11.74	−3.74	13.99	1.19
10	5.60	4.40	19.36	3.46
4	4.26	−0.26	.07	.02

$$\chi^2 = 27.56$$
$$df = (r-1)(c-1) = (3-1)(4-1) = 6$$

As an illustration, consider once again the observed frequencies in Table 16.5:

50	70	120
19	41	60
69	111	180 $= N$

$$\chi^2 = \frac{180(50 \times 41 - 70 \times 19)^2}{(120)(60)(69)(111)}$$
$$= \frac{(180)(720)^2}{(120)(60)(69)(111)}$$
$$= 1.692$$

Note that this is the same value obtained in Table 16.5.

MEASURES OF STRENGTH OF ASSOCIATION IN TWO-VARIABLE TABLES

The χ^2 test of independence allows you to make decisions about *whether* there is a relationship between two variables, using frequency data. Thus, if H_0 is rejected, you conclude that there *is* a statistically significant relationship between the two variables. As we pointed out in Chapter 12, however, statistical significance does not indicate the *strength* of the relationship; a significant result means only that the relationship in the population is unlikely to be zero. As was the case with t (see Chapter 12), it is very desirable to have a measure of the strength of the relationship—that is, have an index of degree of correlation. We will discuss three such measures.

The phi coefficient (ϕ) for 2×2 tables. Suppose that a researcher wants to know if there is a relationship between sex and the type of automobile preferred. He obtains a sample of 50 men and 50 women and asks each individual whether he or she would prefer to own a sports car or a more conventional automobile. The results are shown in Table 16.7.

The computed value of χ^2 of 4 is statistically significant for $df = 1$ using the .05 criterion. Thus, the two variables of sex and automobile preference *are* related; men tend to prefer sports cars

Table 16.7 *Table of frequencies relating sex to type of automobile preferred*

		sex		total people
		female	male	
automobile preference	sports car	20	30	50
	regular car	30	20	50
	Total	50 +	50 =	100 = N

f_e for each cell = 25

$$\chi^2 = \frac{100[(20 \times 20) - (30 \times 30)]^2}{(50)(50)(50)(50)}$$

$$= 4$$

OR

$$\chi^2 = \frac{(20-25)^2}{25} + \frac{(30-25)^2}{25}$$
$$+ \frac{(30-25)^2}{25} + \frac{(20-25)^2}{25}$$

$$= 4$$

and women tend to prefer regular automobiles. But how strong is this relationship? Since a 2 × 2 χ^2 table is involved, the answer is provided by converting χ^2 to a correlation coefficient called the *phi coefficient* (symbolized by ϕ), where

$$\phi = \sqrt{\frac{\chi^2}{N}}$$

Thus,

$$\phi = \sqrt{\frac{4.0}{100}}$$

$$= \sqrt{.04}$$

$$= .20$$

The phi coefficient is interpreted as a Pearson r. In fact, it *is* a Pearson r; you would get the same answer of .20 if you assigned scores of 0 and 1 to sex (for example, $0 =$ female, $1 =$ male) and to automobile preference (for example, $0 =$ prefers sports cars, $1 =$ prefers regular cars) and computed the correlation between the two variables using the usual formula for the Pearson r given in Chapter 11. The absolute value of ϕ varies between 0 and 1; the larger the value of ϕ, the stronger is the relationship between the two variables. When the assignment of score values is arbitrary (as it is in the present case; it would be just as acceptable to score 1 for females and 0 for males, or 1 for prefers sports cars and 0 for prefers regular cars), the sign of ϕ is irrelevant (and is therefore reported without sign); you must look at the data to determine the direction of the findings (in our example, that more men prefer sports cars).

In the present example, you would conclude from the value of ϕ that the relationship between sex and automobile preference in the population, while likely to be greater than zero, is also likely to be fairly weak. In other words, there are likely to be numerous exceptions to the conclusion, as evidenced by the 20 out of 50 women in Table 16.7 who preferred sports cars and the 20 out of 50 men who preferred regular automobiles, and ϕ sums up this information conveniently in a readily interpretable correlation.

Note that there is no need to carry out a special significance test for ϕ. It has already been done with χ^2. In fact, the χ^2 test on a 2 × 2 table can just as well be viewed as testing the null hypothesis that the population ϕ value equals zero. When χ^2 is large, this hypothesis is rejected, which obviously means that there is some degree of relationship other than zero.

The contingency coefficient, C. When the problem is to determine the strength of the relationship between two variables, data are in terms of frequencies, and a table larger than 2×2 is involved, one can compute a *contingency coefficient* (symbolized by C), where

$$C = \sqrt{\frac{\chi^2}{N + \chi^2}}$$

As an illustration, consider the data in Table 16.6:

$$C = \sqrt{\frac{27.56}{300 + 27.56}}$$

$$= \sqrt{.0842}$$

$$= .29$$

This value of C is necessarily statistically significant (that is, different from zero in the population) because the value of χ^2 on which it is based is statistically significant (as was pointed out for ϕ). However, it is fairly small, indicating that the relationship between ethnic group and political preference in the population may well be only a small to moderate one.

Unlike ϕ, however, C is *not* a Pearson r coefficient. Values of C can never be greater than 1.0, but the *maximum* value of the C coefficient in a given situation is strongly influenced by the size of the χ^2 table. It can be shown that the maximum value of C is equal to:

$$C_{max} = \sqrt{\frac{k - 1}{k}}$$

where $k =$ the *smaller* of r (number of rows) or c (number of columns), and either one if $r = c$

Thus, for a 3×4 (or 4×3) table, or a 3×3 table,

$$C_{max} = \sqrt{\frac{3 - 1}{3}}$$

$$= .82$$

In spite of this difficulty, C is a long established and widely used index of relationship where larger than 2×2 χ^2 tables are involved.

Cramér's ϕ. An index of strength of relationship for use with larger than 2×2 tables which is free of the dependence on table size that causes difficulty with the C coefficient is Cramér's ϕ. This is a generalization of the 2×2 ϕ coefficient which is applicable to $r \times c$ tables of frequencies, and always varies between 0 and 1 regardless of the size of the table. It is simply

$$\text{Cramér's } \phi = \sqrt{\frac{\chi^2}{N(k-1)}}$$

where, as before, $k =$ smaller of r or c (or, when $r = c$, $k = r = c$) For the example in Table 16.6, which is a 3×4 table,

$$\text{Cramér's } \phi = \sqrt{\frac{27.56}{300(3-1)}}$$

$$= \sqrt{.0459}$$

$$= .21$$

Both C and Cramér's ϕ are applicable in the same circumstances. The traditionally used measure is C; Cramér's ϕ is a superior index, but not yet widely known among behavioral scientists.

SUMMARY

Chi square is used when the data are frequencies or "head counts." One-variable chi square analysis permits the comparison of a set of obtained frequencies to a null-hypothesized distribution of frequencies. Two-variable chi square permits a test of the independence of the two variables, and *phi* (used with 2×2 tables) and C or *Cramér's* ϕ (used with larger tables) are useful as measures of the strength of the relationship between the two variables.

Table A. *Squares, Square Roots, and Reciprocals of Numbers from 1 to 1000**

N	N^2	$\sqrt{N}$	$1/N$	N	N^2	$\sqrt{N}$	$1/N$
1	1	1.0000	1.000000	41	1681	6.4031	.024390
2	4	1.4142	.500000	42	1764	6.4807	.023810
3	9	1.7321	.333333	43	1849	6.5574	.023256
4	16	2.0000	.250000	44	1936	6.6332	.022727
5	25	2.2361	.200000	45	2025	6.7082	.022222
6	36	2.4495	.166667	46	2116	6.7823	.021739
7	49	2.6458	.142857	47	2209	6.8557	.021277
8	64	2.8284	.125000	48	2304	6.9282	.020833
9	81	3.0000	.111111	49	2401	7.0000	.020408
10	100	3.1623	.100000	50	2500	7.0711	.020000
11	121	3.3166	.090909	51	2601	7.1414	.019608
12	144	3.4641	.083333	52	2704	7.2111	.019231
13	169	3.6056	.076923	53	2809	7.2801	.018868
14	196	3.7417	.071429	54	2916	7.3485	.018519
15	225	3.8730	.066667	55	3025	7.4162	.018182
16	256	4.0000	.062500	56	3136	7.4833	.017857
17	289	4.1231	.058824	57	3249	7.5498	.017544
18	324	4.2426	.055556	58	3364	7.6158	.017241
19	361	4.3589	.052632	59	3481	7.6811	.016949
20	400	4.4721	.050000	60	3600	7.7460	.016667
21	441	4.5826	.047619	61	3721	7.8102	.016393
22	484	4.6904	.045455	62	3844	7.8740	.016129
23	529	4.7958	.043478	63	3969	7.9373	.015873
24	576	4.8990	.041667	64	4096	8.0000	.015625
25	625	5.0000	.040000	65	4225	8.0623	.015385
26	676	5.0990	.038462	66	4356	8.1240	.015152
27	729	5.1962	.037037	67	4489	8.1854	.014925
28	784	5.2915	.035714	68	4624	8.2462	.014706
29	841	5.3852	.034483	69	4761	8.3066	.014493
30	900	5.4772	.033333	70	4900	8.3666	.014286
31	961	5.5678	.032258	71	5041	8.4261	.014085
32	1024	5.6569	.031250	72	5184	8.4853	.013889
33	1089	5.7446	.030303	73	5329	8.5440	.013699
34	1156	5.8310	.029412	74	5476	8.6023	.013514
35	1225	5.9161	.028571	75	5625	8.6603	.013333
36	1296	6.0000	.027778	76	5776	8.7178	.013158
37	1369	6.0828	.027027	77	5929	8.7750	.012987
38	1444	6.1644	.026316	78	6084	8.8318	.012821
39	1521	6.2450	.025641	79	6241	8.8882	.012658
40	1600	6.3246	.025000	80	6400	8.9443	.012500

* Reprinted from A. L. Edwards. "Experimental Design in Psychological Research," Holt, Rinehart, and Winston, New York, 1968. Adapted from original tables from J. W. Dunlap and A. K. Kurtz, "Handbook of Statistical Nomographs, Tables, and Formulas." World Book Co., New York, 1932. Reproduced by permission of the authors.

Table A (*Continued*)

N	N^2	$\sqrt{N}$	$1/N$	N	N^2	$\sqrt{N}$	$1/N$
81	6561	9.0000	.012346	121	14641	11.0000	.00826446
82	6724	9.0554	.012195	122	14884	11.0454	.00819672
83	6889	9.1104	.012048	123	15129	11.0905	.00813008
84	7056	9.1652	.011905	124	15376	11.1355	.00806452
85	7225	9.2195	.011765	125	15625	11.1803	.00800000
86	7396	9.2736	.011628	126	15876	11.2250	.00793651
87	7569	9.3274	.011494	127	16129	11.2694	.00787402
88	7744	9.3808	.011364	128	16384	11.3137	.00781250
89	7921	9.4340	.011236	129	16641	11.3578	.00775194
90	8100	9.4868	.011111	130	16900	11.4018	.00769231
91	8281	9.5394	.010989	131	17161	11.4455	.00763359
92	8464	9.5917	.010870	132	17424	11.4891	.00757576
93	8649	9.6437	.010753	133	17689	11.5326	.00751880
94	8836	9.6954	.010638	134	17956	11.5758	.00746269
95	9025	9.7468	.010526	135	18225	11.6190	.00740741
96	9216	9.7980	.010417	136	18496	11.6619	.00735294
97	9409	9.8489	.010309	137	18769	11.7047	.00729927
98	9604	9.8995	.010204	138	19044	11.7473	.00724638
99	9801	9.9499	.010101	139	19321	11.7898	.00719424
100	10000	10.0000	.010000	140	19600	11.8322	.00714286
101	10201	10.0499	.00990099	141	19881	11.8743	.00709220
102	10404	10.0995	.00980392	142	20164	11.9164	.00704225
103	10609	10.1489	.00970874	143	20449	11.9583	.00699301
104	10816	10.1980	.00961538	144	20736	12.0000	.00694444
105	11025	10.2470	.00952381	145	21025	12.0416	.00689655
106	11236	10.2956	.00943396	146	21316	12.0830	.00684932
107	11449	10.3441	.00934579	147	21609	12.1244	.00680272
108	11664	10.3923	.00925926	148	21904	12.1655	.00675676
109	11881	10.4403	.00917431	149	22201	12.2066	.00671141
110	12100	10.4881	.00909091	150	22500	12.2474	.00666667
111	12321	10.5357	.00900901	151	22801	12.2882	.00662252
112	12544	10.5830	.00892857	152	23104	12.3288	.00657895
113	12769	10.6301	.00884956	153	23409	12.3693	.00653595
114	12996	10.6771	.00877193	154	23716	12.4097	.00649351
115	13225	10.7238	.00869565	155	24025	12.4499	.00645161
116	13456	10.7703	.00862069	156	24336	12.4900	.00641026
117	13689	10.8167	.00854701	157	24649	12.5300	.00636943
118	13924	10.8628	.00847458	158	24964	12.5698	.00632911
119	14161	10.9087	.00840336	159	25281	12.6095	.00628931
120	14400	10.9545	.00833333	160	25600	12.6491	.00625000

Table A (*Continued*)

N	N^2	$\sqrt{N}$	$1/N$	N	N^2	$\sqrt{N}$	$1/N$
161	25921	12.6886	.00621118	201	40401	14.1774	.00497512
162	26244	12.7279	.00617284	202	40804	14.2127	.00495050
163	26569	12.7671	.00613497	203	41209	14.2478	.00492611
164	26896	12.8062	.00609756	204	41616	14.2829	.00490196
165	27225	12.8452	.00606061	205	42025	14.3178	.00487805
166	27556	12.8841	.00602410	206	42436	14.3527	.00485437
167	27889	12.9228	.00598802	207	42849	14.3875	.00483092
168	28224	12.9615	.00595238	208	43264	14.4222	.00480769
169	28561	13.0000	.00591716	209	43681	14.4568	.00478469
170	28900	13.0384	.00588235	210	44100	14.4914	.00476190
171	29241	13.0767	.00584795	211	44521	14.5258	.00473934
172	29584	13.1149	.00581395	212	44944	14.5602	.00471698
173	29929	13.1529	.00578035	213	45369	14.5945	.00469484
174	30276	13.1909	.00574713	214	45796	14.6287	.00467290
175	30625	13.2288	.00571429	215	46225	14.6629	.00465116
176	30976	13.2665	.00568182	216	46656	14.6969	.00462963
177	31329	13.3041	.00564972	217	47089	14.7309	.00460829
178	31684	13.3417	.00561798	218	47524	14.7648	.00458716
179	32041	13.3791	.00558659	219	47961	14.7986	.00456621
180	32400	13.4164	.00555556	220	48400	14.8324	.00454545
181	32761	13.4536	.00552486	221	48841	14.8661	.00452489
182	33124	13.4907	.00549451	222	49284	14.8997	.00450450
183	33489	13.5277	.00546448	223	49729	14.9332	.00448430
184	33856	13.5647	.00543478	224	50176	14.9666	.00446429
185	34225	13.6015	.00540541	225	50625	15.0000	.00444444
186	34596	13.6382	.00537634	226	51076	15.0333	.00442478
187	34969	13.6748	.00534759	227	51529	15.0665	.00440529
188	35344	13.7113	.00531915	228	51984	15.0997	.00438596
189	35721	13.7477	.00529101	229	52441	15.1327	.00436681
190	36100	13.7840	.00526316	230	52900	15.1658	.00434783
191	36481	13.8203	.00523560	231	53361	15.1987	.00432900
192	36864	13.8564	.00520833	232	53824	15.2315	.00431034
193	37249	13.8924	.00518135	233	54289	15.2643	.00429185
194	37636	13.9284	.00515464	234	54756	15.2971	.00427350
195	38025	13.9642	.00512821	235	55225	15.3297	.00425532
196	38416	14.0000	.00510204	236	55696	15.3623	.00423729
197	38809	14.0357	.00507614	237	56169	15.3948	.00421941
198	39204	14.0712	.00505051	238	56644	15.4272	.00420168
199	39601	14.1067	.00502513	239	57121	15.4596	.00418410
200	40000	14.1421	.00500000	240	57600	15.4919	.00416667

Table A (*Continued*)

N	N^2	$\sqrt{N}$	$1/N$	N	N^2	$\sqrt{N}$	$1/N$
241	58081	15.5242	.00414938	281	78961	16.7631	.00355872
242	58564	15.5563	.00413223	282	79524	16.7929	.00354610
243	59049	15.5885	.00411523	283	80089	16.8226	.00353357
244	59536	15.6205	.00409836	284	80656	16.8523	.00352113
245	60025	15.6525	.00408163	285	81225	16.8819	.00350877
246	60516	15.6844	.00406504	286	81796	16.9115	.00349650
247	61009	15.7162	.00404858	287	82369	16.9411	.00348432
248	61504	15.7480	.00403226	288	82944	16.9706	.00347222
249	62001	15.7797	.00401606	289	83521	17.0000	.00346021
250	62500	15.8114	.00400000	290	84100	17 0294	.00344828
251	63001	15.8430	.00398406	291	84681	17.0587	.00343643
252	63504	15.8745	.00396825	292	85264	17.0880	.00342466
253	64009	15.9060	.00395257	293	85849	17.1172	.00341297
254	64516	15.9371	.00393701	294	86436	17.1464	.00340136
255	65025	15.9687	.00392157	295	87025	17.1756	.00338983
256	65536	16.0000	.00390625	296	87616	17.2047	.00337838
257	66049	16.0312	.00389105	297	88209	17.2337	.00336700
258	66564	16.0624	.00387597	298	88804	17.2627	.00335570
259	67081	16.0935	.00386100	299	89401	17.2916	.00334448
260	67600	16.1245	.00384615	300	90000	17.3205	.00333333
261	68121	16.1555	.00383142	301	90601	17.3494	.00332226
262	68644	16.1864	.00381679	302	91204	17.3781	.00331126
263	69169	16.2173	.00380228	303	91809	17.4069	.00330033
264	69696	16.2481	.00378788	304	92416	17.4356	.00328947
265	70225	16.2788	.00377358	305	93025	17.4642	.00327869
266	70756	16.3095	.00375940	306	93636	17.4929	.00326797
267	71289	16.3401	.00374532	307	94249	17.5214	.00325733
268	71824	16.3707	.00373134	308	94864	17.5499	.00324675
269	72361	16.4012	.00371747	309	95481	17.5784	.00323625
270	72900	16.4317	.00370370	310	96100	17.6068	.00322581
271	73441	16.4621	.00369004	311	96721	17.6352	.00321543
272	73984	16.4924	.00367647	312	97344	17.6635	.00320513
273	74529	16.5227	.00366300	313	97969	17.6918	.00319489
274	75076	16.5529	.00364964	314	98596	17.7200	.00318471
275	75625	16.5831	.00363636	315	99225	17.7482	.00317460
276	76176	16.6132	.00362319	316	99856	17.7764	.00316456
277	76729	16.6433	.00361011	317	100489	17.8045	.00315457
278	77284	16.6733	.00359712	318	101124	17.8326	.00314465
279	77841	16.7033	.00358423	319	101761	17.8606	.00313480
280	78400	16.7332	.00357143	320	102400	17.8885	.00312500

Table A (*Continued*)

N	N²	$\sqrt{N}$	1/N	N	N²	$\sqrt{N}$	1/N
321	103041	17.9165	.00311526	361	130321	19.0000	.00277008
322	103684	17.9444	.00310559	362	131044	19.0263	.00276243
323	104329	17.9722	.00309598	363	131769	19.0526	.00275482
324	104976	18.0000	.00308642	364	132496	19.0788	.00274725
325	105625	18.0278	.00307692	365	133225	19.1050	.00273973
326	106276	18.0555	.00306748	366	133956	19.1311	.00273224
327	106929	18.0831	.00305810	367	134689	19.1572	.00272480
328	107584	18.1108	.00304878	368	135424	19.1833	.00271739
329	108241	18.1384	.00303951	369	136161	19.2094	.00271003
330	108900	18.1659	.00303030	370	136900	19.2354	.00270270
331	109561	18.1934	.00302115	371	137641	19.2614	.00269542
332	110224	18.2209	.00301205	372	138384	19.2873	.00268817
333	110889	18.2483	.00300300	373	139129	19.3132	.00268097
334	111556	18.2757	.00299401	374	139876	19.3391	.00267380
335	112225	18.3030	.00298507	375	140625	19.3649	.00266667
336	112896	18.3303	.00297619	376	141376	19.3907	.00265957
337	113569	18.3576	.00296736	377	142129	19.4165	.00265252
338	114244	18.3848	.00295858	378	142884	19.4422	.00264550
339	114921	18.4120	.00294985	379	143641	19.4679	.00263852
340	115600	18.4391	.00294118	380	144400	19.4936	.00263158
341	116281	18.4662	.00293255	381	145161	19.5192	.00262467
342	116964	18.4932	.00292398	382	145924	19.5448	.00261780
343	117649	18.5203	.00291545	383	146689	19.5704	.00261097
344	118336	18.5472	.00290698	384	147456	19.5959	.00260417
345	119025	18.5742	.00289855	385	148225	19.6214	.00259740
346	119716	18.6011	.00289017	386	148996	19.6469	.00259067
347	120409	18.6279	.00288184	387	149769	19.6723	.00258398
348	121104	18.6548	.00287356	388	150544	19.6977	.00257732
349	121801	18.6815	.00286533	389	151321	19.7231	.00257069
350	122500	18.7083	.00285714	390	152100	19.7484	.00256410
351	123201	18.7350	.00284900	391	152881	19.7737	.00255754
352	123904	18.7617	.00284091	392	153664	19.7990	.00255102
353	124609	18.7883	.00283286	393	154449	19.8242	.00254453
354	125316	18.8149	.00282486	394	155236	19.8494	.00253807
355	126025	18.8414	.00281690	395	156025	19.8746	.00253165
356	126736	18.8680	.00280899	396	156816	19.8997	.00252525
357	127449	18.8944	.00280112	397	157609	19.9249	.00251889
358	128164	18.9209	.00279330	398	158404	19.9499	.00251256
359	128881	18.9473	.00278552	399	159201	19.9750	.00250627
360	129600	18.9737	.00277778	400	160000	20.0000	.00250000

Table A (*Continued*)

N	N^2	$\sqrt{N}$	$1/N$	N	N^2	$\sqrt{N}$	$1/N$
401	160801	20.0250	.00249377	441	194481	21.0000	.00226757
402	161604	20.0499	.00248756	442	195364	21.0238	.00226244
403	162409	20.0749	.00248139	443	196249	21.0476	.00225734
404	163216	20.0998	.00247525	444	197136	21.0713	.00225225
405	164025	20.1246	.00246914	445	198025	21.0950	.00224719
406	164836	20.1494	.00246305	446	198916	21.1187	.00224215
407	165649	20.1742	.00245700	447	199809	21.1424	.00223714
408	166464	20.1990	.00245098	448	200704	21.1660	.00223214
409	167281	20.2237	.00244499	449	201601	21.1896	.00222717
410	168100	20.2485	.00243902	450	202500	21.2132	.00222222
411	168921	20.2731	.00243309	451	203401	21.2368	.00221729
412	169744	20.2978	.00242718	452	204304	21.2603	.00221239
413	170569	20.3224	.00242131	453	205209	21.2838	.00220751
414	171396	20.3470	.00241546	454	206116	21.3073	.00220264
415	172225	20.3715	.00240964	455	207025	21.3307	.00219780
416	173056	20.3961	.00240385	456	207936	21.3542	.00219298
417	173889	20.4206	.00239808	457	208849	21.3776	.00218818
418	174724	20.4450	.00239234	458	209764	21.4009	.00218341
419	175561	20.4695	.00238663	459	210681	21.4243	.00217865
420	176400	20.4939	.00238095	460	211600	21.4476	.00217391
421	177241	20.5183	.00237530	461	212521	21.4709	.00216920
422	178084	20.5426	.00236967	462	213444	21.4942	.00216450
423	178929	20.5670	.00236407	463	214369	21.5174	.00215983
424	179776	20.5913	.00235849	464	215296	21.5407	.00215517
425	180625	20.6155	.00235294	465	216225	21.5639	.00215054
426	181476	20.6398	.00234742	466	217156	21.5870	.00214592
427	182329	20.6640	.00234192	467	218089	21.6102	.00214133
428	183184	20.6882	.00233645	468	219024	21.6333	.00213675
429	184041	20.7123	.00233100	469	219961	21.6564	.00213220
430	184900	20.7364	.00232558	470	220900	21.6795	.00212766
431	185761	20.7605	.00232019	471	221841	21.7025	.00212314
432	186624	20.7846	.00231481	472	222784	21.7256	.00211864
433	187489	20.8087	.00230947	473	223729	21.7486	.00211416
434	188356	20.8327	.00230415	474	224676	21.7715	.00210970
435	189225	20.8567	.00229885	475	225625	21.7945	.00210526
436	190096	20.8806	.00229358	476	226576	21.8174	.00210084
437	190969	20.9045	.00228833	477	227529	21.8403	.00209644
438	191844	20.9284	.00228311	478	228484	21.8632	.00209205
439	192721	20.9523	.00227790	479	229441	21.8861	.00208768
440	193600	20.9762	.00227273	480	230400	21.9089	.00208333

Table A (*Continued*)

N	N^2	$\sqrt{N}$	$1/N$	N	N^2	$\sqrt{N}$	$1/N$
481	231361	21.9317	.00207900	521	271441	22.8254	.00191939
482	232324	21.9545	.00207469	522	272484	22.8473	.00191571
483	233289	21.9773	.00207039	523	273529	22.8692	.00191205
484	234256	22.0000	.00206612	524	274576	22.8910	.00190840
485	235225	22.0227	.00206186	525	275625	22.9129	.00190476
486	236196	22.0454	.00205761	526	276676	22.9347	.00190114
487	237169	22.0681	.00205339	527	277729	22.9565	.00189753
488	238144	22.0907	.00204918	528	278784	22.9783	.00189394
489	239121	22.1133	.00204499	529	279841	23.0000	.00189036
490	240100	22.1359	.00204082	530	280900	23.0217	.00188679
491	241081	22.1585	.00203666	531	281961	23.0434	.00188324
492	242064	22.1811	.00203252	532	283024	23.0651	.00187970
493	243049	22.2036	.00202840	533	284089	23.0868	.00187617
494	244036	22.2261	.00202429	534	285156	23.1084	.00187266
495	245025	22.2486	.00202020	535	286225	23.1301	.00186916
496	246016	22.2711	.00201613	536	287296	23.1517	.00186567
497	247009	22.2935	.00201207	537	288369	23.1733	.00186220
498	248004	22.3159	.00200803	538	289444	23.1948	.00185874
499	249001	22.3383	.00200401	539	290521	23.2164	.00185529
500	250000	22.3607	.00200000	540	291600	23.2379	.00185185
501	251001	22.3830	.00199601	541	292681	23.2594	.00184843
502	252004	22.4054	.00199203	542	293764	23.2809	.00184502
503	253009	22.4277	.00198807	543	294849	23.3024	.00184162
504	254016	22.4499	.00198413	544	295936	23.3238	.00183824
505	255025	22.4722	.00198020	545	297025	23.3452	.00183486
506	256036	22.4944	.00197628	546	298116	23.3666	.00183150
507	257049	22.5167	.00197239	547	299209	23.3880	.00182815
508	258064	22.5389	.00196850	548	300304	23.4094	.00182482
509	259081	22.5610	.00196464	549	301401	23.4307	.00182149
510	260100	22.5832	.00196078	550	302500	23.4521	.00181818
511	261121	22.6053	.00195695	551	303601	23.4734	.00181488
512	262144	22.6274	.00195312	552	304704	23.4947	.00181159
513	263169	22.6495	.00194932	553	305809	23.5160	.00180832
514	264196	22.6716	.00194553	554	306916	23.5372	.00180505
515	265225	22.6936	.00194175	555	308025	23.5584	.00180180
516	266256	22.7156	.00193798	556	309136	23.5797	.00179856
517	267289	22.7376	.00193424	557	310249	23.6008	.00179533
518	268324	22.7596	.00193050	558	311364	23.6220	.00179211
519	269361	22.7816	.00192678	559	312481	23.6432	.00178891
520	270400	22.8035	.00192308	560	313600	23.6643	.00178571

Table A (*Continued*)

N	N²	√N	1/N	N	N²	√N	1/N
561	314721	23.6854	.00178253	601	361201	24.5153	.00166389
562	315844	23.7065	.00177936	602	362404	24.5357	.00166113
563	316969	23.7276	.00177620	603	363609	24.5561	.00165837
564	318096	23.7487	.00177305	604	364816	24.5764	.00165563
565	319225	23.7697	.00176991	605	366025	24.5967	.00165289
566	320356	23.7908	.00176678	606	367236	24.6171	.00165017
567	321489	23.8118	.00176367	607	368449	24.6374	.00164745
568	322624	23.8328	.00176056	608	369664	24.6577	.00164474
569	323761	23.8537	.00175747	609	370881	24.6779	.00164204
570	324900	23.8747	.00175439	610	372100	24.6982	.00163934
571	326041	23.8956	.00175131	611	373321	24.7184	.00163666
572	327184	23.9165	.00174825	612	374544	24.7386	.00163399
573	328329	23.9374	.00174520	613	375769	24.7588	.00163132
574	329476	23.9583	.00174216	614	376996	24.7790	.00162866
575	330625	23.9792	.00173913	615	378225	24.7992	.00162602
576	331776	24.0000	.00173611	616	379456	24.8193	.00162338
577	332929	24.0208	.00173310	617	380689	24.8395	.00162075
578	334084	24.0416	.00173010	618	381924	24.8596	.00161812
579	335241	24.0624	.00172712	619	383161	24.8797	.00161551
580	336400	24.0832	.00172414	620	384400	24.8998	.00161290
581	337561	24.1039	.00172117	621	385641	24.9199	.00161031
582	338724	24.1247	.00171821	622	386884	24.9399	.00160772
583	339889	24.1454	.00171527	623	388129	24.9600	.00160514
584	341056	24.1661	.00171233	624	389376	24.9800	.00160256
585	342225	24.1868	.00170940	625	390625	25.0000	.00160000
586	343396	24.2074	.00170648	626	391876	25.0200	.00159744
587	344569	24.2281	.00170358	627	393129	25.0400	.00159490
588	345744	24.2487	.00170068	628	394384	25.0599	.00159236
589	346921	24.2693	.00169779	629	395641	25.0799	.00158983
590	348100	24.2899	.00169492	630	396900	25.0998	.00158730
591	349281	24.3105	.00169205	631	398161	25.1197	.00158479
592	350464	24.3311	.00168919	632	399424	25.1396	.00158228
593	351649	24.3516	.00168634	633	400689	25.1595	.00157978
594	352836	24.3721	.00168350	634	401956	25.1794	.00157729
595	354025	24.3926	.00168067	635	403225	25.1992	.00157480
596	355216	24.4131	.00167785	636	404496	25.2190	.00157233
597	356409	24.4336	.00167504	637	405769	25.2389	.00156986
598	357604	24.4540	.00167224	638	407044	25.2587	.00156740
599	358801	24.4745	.00166945	639	408321	25.2784	.00156495
600	360000	24.4949	.00166667	640	409600	25.2982	.00156250

Table A (*Continued*)

N	N^2	$\sqrt{N}$	$1/N$	N	N^2	$\sqrt{N}$	$1/N$
641	410881	25.3180	.00156006	681	463761	26.0960	.00146843
642	412164	25.3377	.00155763	682	465124	26.1151	.00146628
643	413449	25.3574	.00155521	683	466489	26.1343	.00146413
644	414736	25.3772	.00155280	684	467856	26.1534	.00146199
645	416025	25.3969	.00155039	685	469225	26.1725	.00145985
646	417316	25.4165	.00154799	686	470596	26.1916	.00145773
647	418609	25.4362	.00154560	687	471969	26.2107	.00145560
648	419904	25.4558	.00154321	688	473344	26.2298	.00145349
649	421201	25.4755	.00154083	689	474721	26.2488	.00145138
650	422500	25.4951	.00153846	690	476100	26.2679	.00144928
651	423801	25.5147	.00153610	691	477481	26.2869	.00144718
652	425104	25.5343	.00153374	692	478864	26.3059	.00144509
653	426409	25.5539	.00153139	693	480249	26.3249	.00144300
654	427716	25.5734	.00152905	694	481636	26.3439	.00144092
655	429025	25.5930	.00152672	695	483025	26.3629	.00143885
656	430336	25.6125	.00152439	696	484416	26.3818	.00143678
657	431649	25.6320	.00152207	697	485809	26.4008	.00143472
658	432964	25.6515	.00151976	698	487204	26.4197	.00143266
659	434281	25.6710	.00151745	699	488601	26.4386	.00143062
660	435600	25.6905	.00151515	700	490000	26.4575	.00142857
661	436921	25.7099	.00151286	701	491401	26.4764	.00142653
662	438244	25.7294	.00151057	702	492804	26.4953	.00142450
663	439569	25.7488	.00150830	703	494209	26.5141	.00142248
664	440896	25.7682	.00150602	704	495616	26.5330	.00142045
665	442225	25.7876	.00150376	705	497025	26.5518	.00141844
666	443556	25.8070	.00150150	706	498436	26.5707	.00141643
667	444889	25.8263	.00149925	707	499849	26.5895	.00141443
668	446224	25.8457	.00149701	708	501264	26.6083	.00141243
669	447561	25.8650	.00149477	709	502681	26.6271	.00141044
670	448900	25.8844	.00149254	710	504100	26.6458	.00140845
671	450241	25.9037	.00149031	711	505521	26.6646	.00140647
672	451584	25.9230	.00148810	712	506944	26.6833	.00140449
673	452929	25.9422	.00148588	713	508369	26.7021	.00140252
674	454276	25.9615	.00148368	714	509796	26.7208	.00140056
675	455625	25.9808	.00148148	715	511225	26.7395	.00139860
676	456976	26.0000	.00147929	716	512656	26.7582	.00139665
677	458329	26.0192	.00147710	717	514089	26.7769	.00139470
678	459684	26.0384	.00147493	718	515524	26.7955	.00139276
679	461041	26.0576	.00147275	719	516961	26.8142	.00139082
680	462400	26.0768	.00147059	720	518400	26.8328	.00138889

Table A (*Continued*)

N	N^2	$\sqrt{N}$	$1/N$	N	N^2	$\sqrt{N}$	$1/N$
721	519841	26.8514	.00138696	761	579121	27.5862	.00131406
722	521284	26.8701	.00138504	762	580644	27.6043	.00131234
723	522729	26.8887	.00138313	763	582169	27.6225	.00131062
724	524176	26.9072	.00138122	764	583696	27.6405	.00130890
725	525625	26.9258	.00137931	765	585225	27.6586	.00130719
726	527076	26.9444	.00137741	766	586756	27.6767	.00130548
727	528529	26.9629	.00137552	767	588289	27.6948	.00130378
728	529984	26.9815	.00137363	768	589824	27.7128	.00130208
729	531441	27.0000	.00137174	769	591361	27.7308	.00130039
730	532900	27.0185	.00136986	770	592900	27.7489	.00129870
731	534361	27.0370	.00136799	771	594441	27.7669	.00129702
732	535824	27.0555	.00136612	772	595984	27.7849	.00129534
733	537289	27.0740	.00136426	773	597529	27.8029	.00129366
734	538756	27.0924	.00136240	774	599076	27.8209	.00129199
735	540225	27.1109	.00136054	775	600625	27.8388	.00129032
736	541696	27.1293	.00135870	776	602176	27.8568	.00128866
737	543169	27.1477	.00135685	777	603729	27.8747	.00128700
738	544644	27.1662	.00135501	778	605284	27.8927	.00128535
739	546121	27.1846	.00135318	779	606841	27.9106	.00128370
740	547600	27.2029	.00135135	780	608400	27.9285	.00128205
741	549081	27.2213	.00134953	781	609961	27.9464	.00128041
742	550564	27.2397	.00134771	782	611524	27.9643	.00127877
743	552049	27.2580	.00134590	783	613089	27.9821	.00127714
744	553536	27.2764	.00134409	784	614656	28.0000	.00127551
745	555025	27.2947	.00134228	785	616225	28.0179	.00127389
746	556516	27.3130	.00134048	786	617796	28.0357	.00127226
747	558009	27.3313	.00133869	787	619369	28.0535	.00127065
748	559504	27.3496	.00133690	788	620944	28.0713	.00126904
749	561001	27.3679	.00133511	789	622521	28.0891	.00126743
750	562500	27.3861	.00133333	790	624100	28.1069	.00126582
751	564001	27.4044	.00133156	791	625681	28.1247	.00126422
752	565504	27.4226	.00132979	792	627264	28.1425	.00126263
753	567009	27.4408	.00132802	793	628849	28.1603	.00126103
754	568516	27.4591	.00132626	794	630436	28.1780	.00125945
755	570025	27.4773	.00132450	795	632025	28.1957	.00125786
756	571536	27.4955	.00132275	796	633616	28.2135	.00125628
757	573049	27.5136	.00132100	797	635209	28.2312	.00125471
758	574564	27.5318	.00131926	798	636804	28.2489	.00125313
759	576081	27.5500	.00131752	799	638401	28.2666	.00125156
760	577600	27.5681	.00131579	800	640000	28.2843	.00125000

Table A (*Continued*)

N	N^2	$\sqrt{N}$	$1/N$	N	N^2	$\sqrt{N}$	$1/N$
801	641601	28.3019	.00124844	841	707281	29.0000	.00118906
802	643204	28.3196	.00124688	842	708964	29.0172	.00118765
803	644809	28.3373	.00124533	843	710649	29.0345	.00118624
804	646416	28.3549	.00124378	844	712336	29.0517	.00118483
805	648025	28.3725	.00124224	845	714025	29.0689	.00118343
806	649636	28.3901	.00124069	846	715716	29.0861	.00118203
807	651249	28.4077	.00123916	847	717409	29.1033	.00118064
808	652864	28.4253	.00123762	848	719104	29.1204	.00117925
809	654481	28.4429	.00123609	849	720801	29.1376	.00117786
810	656100	28.4605	.00123457	850	722500	29.1548	.00117647
811	657721	28.4781	.00123305	851	724201	29.1719	.00117509
812	659344	28.4956	.00123153	852	725904	29.1890	.00117371
813	660969	28.5132	.00123001	853	727609	29.2062	.00117233
814	662596	28.5307	.00122850	854	729316	29.2233	.00117096
815	664225	28.5482	.00122699	855	731025	29.2404	.00116959
816	665856	28.5657	.00122549	856	732736	29.2575	.00116822
817	667489	28.5832	.00122399	857	734449	29.2746	.00116686
818	669124	28.6007	.00122249	858	736164	29.2916	.00116550
819	670761	28.6182	.00122100	859	737881	29.3087	.00116414
820	672400	28.6356	.00121951	860	739600	29.3258	.00116279
821	674041	28.6531	.00121803	861	741321	29.3428	.00116144
822	675684	28.6705	.00121655	862	743044	29.3598	.00116009
823	677329	28.6880	.00121507	863	744769	29.3769	.00115875
824	678976	28.7054	.00121359	864	746496	29.3939	.00115741
825	680625	28.7228	.00121212	865	748225	29.4109	.00115607
826	682276	28.7402	.00121065	866	749956	29.4279	.00115473
827	683929	28.7576	.00120919	867	751689	29.4449	.00115340
828	685584	28.7750	.00120773	868	753424	29.4618	.00115207
829	687241	28.7924	.00120627	869	755161	29.4788	.00115075
830	688900	28.8097	.00120482	870	756900	29.4958	.00114943
831	690561	28.8271	.00120337	871	758641	29.5127	.00114811
832	692224	28.8444	.00120192	872	760384	29.5296	.00114679
833	693889	28.8617	.00120048	873	762129	29.5466	.00114548
834	695556	28.8791	.00119904	874	763876	29.5635	.00114416
835	697225	28.8964	.00119760	875	765625	29.5804	.00114286
836	698896	28.9137	.00119617	876	767376	29.5973	.00114155
837	700569	28.9310	.00119474	877	769129	29.6142	.00114025
838	702244	28.9482	.00119332	878	770884	29.6311	.00113895
839	703921	28.9655	.00119190	879	772641	29.6479	.00113766
840	705600	28.9828	.00119048	880	774400	29.6648	.00113636

Table A (*Continued*)

N	N^2	$\sqrt{N}$	$1/N$	N	N^2	$\sqrt{N}$	$1/N$
881	776161	29.6816	.00113507	921	848241	30.3480	.00108578
882	777924	29.6985	.00113379	922	850084	30.3645	.00108460
883	779689	29.7153	.00113250	923	851929	30.3809	.00108342
884	781456	29.7321	.00113122	924	853776	30.3974	.00108225
885	783225	29.7489	.00112994	925	855625	30.4138	.00108108
886	784996	29.7658	.00112867	926	857476	30.4302	.00107991
887	786769	29.7825	.00112740	927	859329	30.4467	.00107875
888	788544	29.7993	.00112613	928	861184	30.4631	.00107759
889	790321	29.8161	.00112486	929	863041	30.4795	.00107643
890	792100	29.8329	.00112360	930	864900	30.4959	.00107527
891	793881	29.8496	.00112233	931	866761	30.5123	.00107411
892	795664	29.8664	.00112108	932	868624	30.5287	.00107296
893	797449	29.8831	.00111982	933	870489	30.5450	.00107181
894	799236	29.8998	.00111857	934	872356	30.5614	.00107066
895	801025	29.9166	.00111732	935	874225	30.5778	.00106952
896	802816	29.9333	.00111607	936	876096	30.5941	.00106838
897	804609	29.9500	.00111483	937	877969	30.6105	.00106724
898	806404	29.9666	.00111359	938	879844	30.6268	.00106610
899	808201	29.9833	.00111235	939	881721	30.6431	.00106496
900	810000	30.0000	.00111111	940	883600	30.6594	.00106383
901	811801	30.0167	.00110988	941	885481	30.6757	.00106270
902	813604	30.0333	.00110865	942	887364	30.6920	.00106157
903	815409	30.0500	.00110742	943	889249	30.7083	.00106045
904	817216	30.0666	.00110619	944	891136	30.7246	.00105932
905	819025	30.0832	.00110497	945	893025	30.7409	.00105820
906	820836	30.0998	.00110375	946	894916	30.7571	.00105708
907	822649	30.1164	.00110254	947	896809	30.7734	.00105597
908	824464	30.1330	.00110132	948	898704	30.7896	.00105485
909	826281	30.1496	.00110011	949	900601	30.8058	.00105374
910	828100	30.1662	.00109890	950	902500	30.8221	.00105263
911	829921	30.1828	.00109769	951	904401	30.8383	.00105152
912	831744	30.1993	.00109649	952	906304	30.8545	.00105042
913	833569	30.2159	.00109529	953	908209	30.8707	.00104932
914	835396	30.2324	.00109409	954	910116	30.8869	.00104822
915	837225	30.2490	.00109290	955	912025	30.9031	.00104712
916	839056	30.2655	.00109170	956	913936	30.9192	.00104603
917	840889	30.2820	.00109051	957	915849	30.9354	.00104493
918	842724	30.2985	.00108932	958	917764	30.9516	.00104384
919	844561	30.3150	.00108814	959	919681	30.9677	.00104275
920	846400	30.3315	.00108696	960	921600	30.9839	.00104167

Table A (*Continued*)

N	N^2	$\sqrt{N}$	$1/N$	N	N^2	$\sqrt{N}$	$1/N$
961	923521	31.0000	.00104058	981	962361	31.3209	.00101937
962	925444	31.0161	.00103950	982	964324	31.3369	.00101833
963	927369	31.0322	.00103842	983	966289	31.3528	.00101729
964	929296	31.0483	.00103734	984	968256	31.3688	.00101626
965	931225	31.0644	.00103627	985	970225	31.3847	.00101523
966	933156	31.0805	.00103520	986	972196	31.4006	.00101420
967	935089	31.0966	.00103413	987	974169	31.4166	.00101317
968	937024	31.1127	.00103306	988	976144	31.4325	.00101215
969	938961	31.1288	.00103199	989	978121	31.4484	.00101112
970	940900	31.1448	.00103093	990	980100	31.4643	.00101010
971	942841	31.1609	.00102987	991	982081	31.4802	.00100908
972	944784	31.1769	.00102881	992	984064	31.4960	.00100806
973	946729	31.1929	.00102775	993	986049	31.5119	.00100705
974	948676	31.2090	.00102669	994	988036	31.5278	.00100604
975	950625	31.2250	.00102564	995	990025	31.5436	.00100503
976	952576	31.2410	.00102459	996	992016	31.5595	.00100402
977	954529	31.2570	.00102354	997	994009	31.5753	.00100301
978	956484	31.2730	.00102249	998	996004	31.5911	.00100200
979	958441	31.2890	.00102145	999	998001	31.6070	.00100100
980	960400	31.3050	.00102041	1000	1000000	31.6228	.00100000

Table B. *Percent area under the normal curve between the mean and z**

z	.00	.01	.02	.03	.04	.05	.06	.07	.08	.09
0.0	00.00	00.40	00.80	01.20	01.60	01.99	02.39	02.79	03.19	03.59
0.1	03.98	04.38	04.78	05.17	05.57	05.96	06.36	06.75	07.14	07.53
0.2	07.93	08.32	08.71	09.10	09.48	09.87	10.26	10.64	11.03	11.41
0.3	11.79	12.17	12.55	12.93	13.31	13.68	14.06	14.43	14.80	15.17
0.4	15.54	15.91	16.28	16.64	17.00	17.36	17.72	18.08	18.44	18.79
0.5	19.15	19.50	19.85	20.19	20.54	20.88	21.23	21.57	21.90	22.24
0.6	22.57	22.91	23.24	23.57	23.89	24.22	24.54	24.86	25.17	25.49
0.7	25.80	26.11	26.42	26.73	27.04	27.34	27.64	27.94	28.23	28.52
0.8	28.81	29.10	29.39	29.67	29.95	30.23	30.51	30.78	31.06	31.33
0.9	31.59	31.86	32.12	32.38	32.64	32.89	33.15	33.40	33.65	33.89
1.0	34.13	34.38	34.61	34.85	35.08	35.31	35.54	35.77	35.99	36.21
1.1	36.43	36.65	36.86	37.08	37.29	37.49	37.70	37.90	38.10	38.30
1.2	38.49	38.69	38.88	39.07	39.25	39.44	39.62	39.80	39.97	40.15
1.3	40.32	40.49	40.66	40.82	40.99	41.15	41.31	41.47	41.62	41.77
1.4	41.92	42.07	42.22	42.36	42.51	42.65	42.79	42.92	43.06	43.19
1.5	43.32	43.45	43.57	43.70	43.82	43.94	44.06	44.18	44.29	44.41
1.6	44.52	44.63	44.74	44.84	44.95	45.05	45.15	45.25	45.35	45.45
1.7	45.54	45.64	45.73	45.82	45.91	45.99	46.08	46.16	46.25	46.33
1.8	46.41	46.49	46.56	46.64	46.71	46.78	46.86	46.93	46.99	47.06
1.9	47.13	47.19	47.26	47.32	47.38	47.44	47.50	47.56	47.61	47.67
2.0	47.72	47.78	47.83	47.88	47.93	47.98	48.03	48.08	48.12	48.17
2.1	48.21	48.26	48.30	48.34	48.38	48.42	48.46	48.50	48.54	48.57
2.2	48.61	48.64	48.68	48.71	48.75	48.78	48.81	48.84	48.87	48.90
2.3	48.93	48.96	48.98	49.01	49.04	49.06	49.09	49.11	49.13	49.16
2.4	49.18	49.20	49.22	49.25	49.27	49.29	49.31	49.32	49.34	49.36
2.5	49.38	49.40	49.41	49.43	49.45	49.46	49.48	49.49	49.51	49.52
2.6	49.53	49.55	49.56	49.57	49.59	49.60	49.61	49.62	49.63	49.64
2.7	49.65	49.66	49.67	49.68	49.69	49.70	49.71	49.72	49.73	49.74
2.8	49.74	49.75	49.76	49.77	49.77	49.78	49.79	49.79	49.80	49.81
2.9	49.81	49.82	49.82	49.83	49.84	49.84	49.85	49.85	49.86	49.86
3.0	49.87									
3.5	49.98									
4.0	49.997									
5.0	49.99997									

* From B. L. Lindquist, "A First Course in Statistics," 2nd ed. Houghton Mifflin, New York, 1942. Reproduced by permission.

Table C. *Critical values of t* *

	Level of significance for one-tailed test					
	.10	.05	.025	.01	.005	.0005
	Level of significance for two-tailed test					
df	.20	.10	.05	.02	.01	.001
1	3.078	6.314	12.706	31.821	63.657	636.619
2	1.886	2.920	4.303	6.965	9.925	31.598
3	1.638	2.353	3.182	4.541	5.841	12.941
4	1.533	2.132	2.776	3.747	4.604	8.610
5	1.476	2.015	2.571	3.365	4.032	6.859
6	1.440	1.943	2.447	3.143	3.707	5.959
7	1.415	1.895	2.365	2.998	3.449	5.405
8	1.397	1.860	2.306	2.896	3.355	5.041
9	1.383	1.833	2.262	2.821	3.250	4.781
10	1.372	1.812	2.228	2.764	3.169	4.587
11	1.363	1.796	2.201	2.718	3.106	4.437
12	1.356	1.782	2.179	2.681	3.055	4.318
13	1.350	1.771	2.160	2.650	3.012	4.221
14	1.345	1.761	2.145	2.624	2.977	4.140
15	1.341	1.753	2.131	2.602	2.947	4.073
16	1.337	1.746	2.120	2.583	2.921	4.015
17	1.333	1.740	2.110	2.567	2.898	3.965
18	1.330	1.734	2.101	2.552	2.878	3.922
19	1.328	1.729	2.093	2.539	2.861	3.883
20	1.325	1.725	2.086	2.528	2.845	3.850
21	1.323	1.721	2.080	2.518	2.831	3.819
22	1.321	1.717	2.074	2.508	2.819	3.792
23	1.319	1.714	2.069	2.500	2.807	3.767
24	1.318	1.711	2.064	2.492	2.797	3.745
25	1.316	1.708	2.060	2.485	2.787	3.725
26	1.315	1.706	2.056	2.479	2.779	3.707
27	1.314	1.703	2.052	2.473	2.771	3.690
28	1.313	1.701	2.048	2.467	2.763	3.674
29	1.311	1.699	2.045	2.462	2.756	3.659
30	1.310	1.697	2.042	2.457	2.750	3.646
40	1.303	1.684	2.021	2.423	2.704	3.551
60	1.296	1.671	2.000	2.390	2.660	3.460
120	1.289	1.658	1.980	2.358	2.617	3.373
∞	1.282	1.645	1.960	2.326	2.576	3.291

* Table C is taken from Table III of R. A. Fisher and F. Yates, "Statistical Tables for Biological, Agricultural and Medical Research," 6th ed. Oliver and Boyd, Edinburgh, 1963. Reproduced by permission of the authors and publishers.

Table D. *Critical values of the Pearson r* *

df (=N−2; N = number of pairs)	Level of significance for one-tailed test			
	.05	.025	.01	.005
	Level of significance for two-tailed test			
	.10	.05	.02	.01
1	.988	.997	.9995	.9999
2	.900	.950	.980	.990
3	.805	.878	.934	.959
4	.729	.811	.882	.917
5	.669	.754	.833	.874
6	.622	.707	.789	.834
7	.582	.666	.750	.798
8	.549	.632	.716	.765
9	.521	.602	.685	.735
10	.497	.576	.658	.708
11	.476	.553	.634	.684
12	.458	.532	.612	.661
13	.441	.514	.592	.641
14	.426	.497	.574	.623
15	.412	.482	.558	.606
16	.400	.468	.542	.590
17	.389	.456	.528	.575
18	.378	.444	.516	.561
19	.369	.433	.503	.549
20	.360	.423	.492	.537
21	.352	.413	.482	.526
22	.344	.404	.472	.515
23	.337	.396	.462	.505
24	.330	.388	.453	.496
25	.323	.381	.445	.487
26	.317	.374	.437	.479
27	.311	.367	.430	.471
28	.306	.361	.423	.463
29	.301	.355	.416	.456
30	.296	.349	.409	.449
35	.275	.325	.381	.418
40	.257	.304	.358	.393
45	.243	.288	.338	.372
50	.231	.273	.322	.354
60	.211	.250	.295	.325
70	.195	.232	.274	.302
80	.183	.217	.256	.283
90	.173	.205	.242	.267
100	.164	.195	.230	.254

* From R. A. Fisher and F. Yates, "Statistical Tables for Biological, Agricultural and Medical Research," 6th ed. Oliver and Boyd, Edinburgh, 1963. Reproduced by permission of authors and publishers.

Table E. *Critical values of r_s (Spearman rank-order correlation coefficient)* *

	Level of significance for one-tailed test			
	.05	.025	.01	.005
	Level of significance for two-tailed test			
No. of pairs (N)	.10	.05	.02	.01
5	.900	1.000	1.000	—
6	.829	.886	.943	1.000
7	.714	.786	.893	.929
8	.643	.738	.833	.881
9	.600	.683	.783	.833
10	.564	.648	.746	.794
12	.506	.591	.712	.777
14	.456	.544	.645	.715
16	.425	.506	.601	.665
18	.399	.475	.564	.625
20	.377	.450	.534	.591
22	.359	.428	.508	.562
24	.343	.409	.485	.537
26	.329	.392	.465	.515
28	.317	.377	.448	.496
30	.306	.364	.432	.478

*From E. G. Olds, *Ann. Math. Statistics* **9** (1938); **20** (1949). Reproduced by permission of publisher.

Table F. *Critical values of F (a = .05 in lightface type, a = .01 in boldface type) **

n₁ degrees of freedom (for greater mean square)

n_2	1	2	3	4	5	6	7	8	9	10	11	12	14	16	20	24	30	40	50	75	100	200	500	∞
1	161 / **4,052**	200 / **4,999**	216 / **5,403**	225 / **5,625**	230 / **5,764**	234 / **5,859**	237 / **5,928**	239 / **5,981**	241 / **6,022**	242 / **6,056**	243 / **6,082**	244 / **6,106**	245 / **6,142**	246 / **6,169**	248 / **6,208**	249 / **6,234**	250 / **6,258**	251 / **6,286**	252 / **6,302**	253 / **6,323**	253 / **6,334**	254 / **6,352**	254 / **6,361**	254 / **6,366**
2	18.51 / **98.49**	19.00 / **99.00**	19.16 / **99.17**	19.25 / **99.25**	19.30 / **99.33**	19.33 / **99.33**	19.36 / **99.34**	19.37 / **99.36**	19.38 / **99.38**	19.39 / **99.40**	19.40 / **99.41**	19.41 / **99.42**	19.42 / **99.43**	19.43 / **99.44**	19.44 / **99.45**	19.45 / **99.46**	19.46 / **99.47**	19.47 / **99.48**	19.47 / **99.48**	19.48 / **99.49**	19.49 / **99.49**	19.49 / **99.49**	19.50 / **99.50**	19.50 / **99.50**
3	10.13 / **34.12**	9.55 / **30.82**	9.28 / **29.46**	9.12 / **28.71**	9.01 / **28.24**	8.94 / **27.91**	8.88 / **27.67**	8.84 / **27.49**	8.81 / **27.34**	8.78 / **27.23**	8.76 / **27.13**	8.74 / **27.05**	8.71 / **26.92**	8.69 / **26.83**	8.66 / **26.69**	8.64 / **26.60**	8.62 / **26.50**	8.60 / **26.41**	8.58 / **26.35**	8.57 / **26.27**	8.56 / **26.23**	8.54 / **26.18**	8.54 / **26.14**	8.53 / **26.12**
4	7.71 / **21.20**	6.94 / **18.00**	6.59 / **16.69**	6.39 / **15.98**	6.26 / **15.52**	6.16 / **15.21**	6.09 / **14.98**	6.04 / **14.80**	6.00 / **14.66**	5.96 / **14.54**	5.93 / **14.45**	5.91 / **14.37**	5.87 / **14.24**	5.84 / **14.15**	5.80 / **14.02**	5.77 / **13.93**	5.74 / **13.83**	5.71 / **13.74**	5.70 / **13.69**	5.68 / **13.61**	5.66 / **13.57**	5.65 / **13.52**	5.64 / **13.48**	5.63 / **13.46**
5	6.61 / **16.26**	5.79 / **13.27**	5.41 / **12.06**	5.19 / **11.39**	5.05 / **10.97**	4.95 / **10.67**	4.88 / **10.45**	4.82 / **10.27**	4.78 / **10.15**	4.74 / **10.05**	4.70 / **9.96**	4.68 / **9.89**	4.64 / **9.77**	4.60 / **9.68**	4.56 / **9.55**	4.53 / **9.47**	4.50 / **9.38**	4.46 / **9.29**	4.44 / **9.24**	4.42 / **9.17**	4.40 / **9.13**	4.38 / **9.07**	4.37 / **9.04**	4.36 / **9.02**
6	5.99 / **13.74**	5.14 / **10.92**	4.76 / **9.78**	4.53 / **9.15**	4.39 / **8.75**	4.28 / **8.47**	4.21 / **8.26**	4.15 / **8.10**	4.10 / **7.98**	4.06 / **7.87**	4.03 / **7.79**	4.00 / **7.72**	3.96 / **7.60**	3.92 / **7.52**	3.87 / **7.39**	3.84 / **7.31**	3.81 / **7.23**	3.77 / **7.14**	3.75 / **7.09**	3.72 / **7.02**	3.71 / **6.99**	3.69 / **6.94**	3.68 / **6.90**	3.67 / **6.88**
7	5.59 / **12.25**	4.74 / **9.55**	4.35 / **8.45**	4.12 / **7.85**	3.97 / **7.46**	3.87 / **7.19**	3.79 / **7.00**	3.73 / **6.84**	3.68 / **6.71**	3.63 / **6.62**	3.60 / **6.54**	3.57 / **6.47**	3.52 / **6.35**	3.49 / **6.27**	3.44 / **6.15**	3.41 / **6.07**	3.38 / **5.98**	3.34 / **5.90**	3.32 / **5.85**	3.29 / **5.78**	3.28 / **5.75**	3.25 / **5.70**	3.24 / **5.67**	3.23 / **5.65**
8	5.32 / **11.26**	4.46 / **8.65**	4.07 / **7.59**	3.84 / **7.01**	3.69 / **6.63**	3.58 / **6.37**	3.50 / **6.19**	3.44 / **6.03**	3.39 / **5.91**	3.34 / **5.82**	3.31 / **5.74**	3.28 / **5.67**	3.23 / **5.56**	3.20 / **5.48**	3.15 / **5.36**	3.12 / **5.28**	3.08 / **5.20**	3.05 / **5.11**	3.03 / **5.06**	3.00 / **5.00**	2.98 / **4.96**	2.96 / **4.91**	2.94 / **4.88**	2.93 / **4.86**
9	5.12 / **10.56**	4.26 / **8.02**	3.86 / **6.99**	3.63 / **6.42**	3.48 / **6.06**	3.37 / **5.80**	3.29 / **5.62**	3.23 / **5.47**	3.18 / **5.35**	3.13 / **5.26**	3.10 / **5.18**	3.07 / **5.11**	3.02 / **5.00**	2.98 / **4.92**	2.93 / **4.80**	2.90 / **4.73**	2.86 / **4.64**	2.82 / **4.56**	2.80 / **4.51**	2.77 / **4.45**	2.76 / **4.41**	2.73 / **4.36**	2.72 / **4.33**	2.71 / **4.31**
10	4.96 / **10.04**	4.10 / **7.56**	3.71 / **6.55**	3.48 / **5.99**	3.33 / **5.64**	3.22 / **5.39**	3.14 / **5.21**	3.07 / **5.06**	3.02 / **4.95**	2.97 / **4.85**	2.94 / **4.78**	2.91 / **4.71**	2.86 / **4.60**	2.82 / **4.52**	2.77 / **4.41**	2.74 / **4.33**	2.70 / **4.25**	2.67 / **4.17**	2.64 / **4.12**	2.61 / **4.05**	2.59 / **4.01**	2.56 / **3.96**	2.55 / **3.93**	2.54 / **3.91**
11	4.84 / **9.65**	3.98 / **7.20**	3.59 / **6.22**	3.36 / **5.67**	3.20 / **5.32**	3.09 / **5.07**	3.01 / **4.88**	2.95 / **4.74**	2.90 / **4.63**	2.86 / **4.54**	2.82 / **4.46**	2.79 / **4.40**	2.74 / **4.29**	2.70 / **4.21**	2.65 / **4.10**	2.61 / **4.02**	2.57 / **3.94**	2.53 / **3.86**	2.50 / **3.80**	2.47 / **3.74**	2.45 / **3.70**	2.42 / **3.66**	2.41 / **3.62**	2.40 / **3.60**
12	4.75 / **9.33**	3.88 / **6.93**	3.49 / **5.95**	3.26 / **5.41**	3.11 / **5.06**	3.00 / **4.82**	2.92 / **4.65**	2.85 / **4.50**	2.80 / **4.39**	2.76 / **4.30**	2.72 / **4.22**	2.69 / **4.16**	2.64 / **4.05**	2.60 / **3.98**	2.54 / **3.86**	2.50 / **3.78**	2.46 / **3.70**	2.42 / **3.61**	2.40 / **3.56**	2.36 / **3.49**	2.35 / **3.46**	2.32 / **3.41**	2.31 / **3.38**	2.30 / **3.36**
13	4.67 / **9.07**	3.80 / **6.70**	3.41 / **5.74**	3.18 / **5.20**	3.02 / **4.86**	2.92 / **4.62**	2.84 / **4.44**	2.77 / **4.30**	2.72 / **4.19**	2.67 / **4.10**	2.63 / **4.02**	2.60 / **3.96**	2.55 / **3.85**	2.51 / **3.78**	2.46 / **3.67**	2.42 / **3.59**	2.38 / **3.51**	2.34 / **3.42**	2.32 / **3.37**	2.28 / **3.30**	2.26 / **3.27**	2.24 / **3.21**	2.22 / **3.18**	2.21 / **3.16**

* Reprinted by permission from "Statistical Methods," 6th ed., by G. W. Snedecor and W. C. Cochran, © 1967 by the Iowa State University Press, Ames, Iowa.

Table F (Continued)

n_1 degrees of freedom (for greater mean square)

n_2	1	2	3	4	5	6	7	8	9	10	11	12	14	16	20	24	30	40	50	75	100	200	500	∞
14	4.60 / 8.86	3.74 / 6.51	3.34 / 5.56	3.11 / 5.03	2.96 / 4.69	2.85 / 4.46	2.77 / 4.28	2.70 / 4.14	2.65 / 4.03	2.60 / 3.94	2.56 / 3.86	2.53 / 3.80	2.48 / 3.70	2.44 / 3.62	2.39 / 3.51	2.35 / 3.43	2.31 / 3.34	2.27 / 3.26	2.24 / 3.21	2.21 / 3.14	2.19 / 3.11	2.16 / 3.06	2.14 / 3.02	2.13 / 3.00
15	4.54 / 8.68	3.68 / 6.36	3.29 / 5.42	3.06 / 4.89	2.90 / 4.56	2.79 / 4.32	2.70 / 4.14	2.64 / 4.00	2.59 / 3.89	2.55 / 3.80	2.51 / 3.73	2.48 / 3.67	2.43 / 3.56	2.39 / 3.48	2.33 / 3.36	2.29 / 3.29	2.25 / 3.20	2.21 / 3.12	2.18 / 3.07	2.15 / 3.00	2.12 / 2.97	2.10 / 2.92	2.08 / 2.89	2.07 / 2.87
16	4.49 / 8.53	3.63 / 6.23	3.24 / 5.29	3.01 / 4.77	2.85 / 4.44	2.74 / 4.20	2.66 / 4.03	2.59 / 3.89	2.54 / 3.78	2.49 / 3.69	2.45 / 3.61	2.42 / 3.55	2.37 / 3.45	2.33 / 3.37	2.28 / 3.25	2.24 / 3.18	2.20 / 3.10	2.16 / 3.01	2.13 / 2.96	2.09 / 2.89	2.07 / 2.86	2.04 / 2.80	2.02 / 2.77	2.01 / 2.75
17	4.45 / 8.40	3.59 / 6.11	3.20 / 5.18	2.96 / 4.67	2.81 / 4.34	2.70 / 4.10	2.62 / 3.93	2.55 / 3.79	2.50 / 3.68	2.45 / 3.59	2.41 / 3.52	2.38 / 3.45	2.33 / 3.35	2.29 / 3.27	2.23 / 3.16	2.19 / 3.08	2.15 / 3.00	2.11 / 2.92	2.08 / 2.86	2.04 / 2.79	2.02 / 2.76	1.99 / 2.70	1.97 / 2.67	1.96 / 2.65
18	4.41 / 8.28	3.55 / 6.01	3.16 / 5.09	2.93 / 4.58	2.77 / 4.25	2.66 / 4.01	2.58 / 3.85	2.51 / 3.71	2.46 / 3.60	2.41 / 3.51	2.37 / 3.44	2.34 / 3.37	2.29 / 3.27	2.25 / 3.19	2.19 / 3.07	2.15 / 3.00	2.11 / 2.91	2.07 / 2.83	2.04 / 2.78	2.00 / 2.71	1.98 / 2.68	1.95 / 2.62	1.93 / 2.59	1.92 / 2.57
19	4.38 / 8.18	3.52 / 5.93	3.13 / 5.01	2.90 / 4.50	2.74 / 4.17	2.63 / 3.94	2.55 / 3.77	2.48 / 3.63	2.43 / 3.52	2.38 / 3.43	2.34 / 3.36	2.31 / 3.30	2.26 / 3.19	2.21 / 3.12	2.15 / 3.00	2.11 / 2.92	2.07 / 2.84	2.02 / 2.76	2.00 / 2.70	1.96 / 2.63	1.94 / 2.60	1.91 / 2.54	1.90 / 2.51	1.88 / 2.49
20	4.35 / 8.10	3.49 / 5.85	3.10 / 4.94	2.87 / 4.43	2.71 / 4.10	2.60 / 3.87	2.52 / 3.71	2.45 / 3.56	2.40 / 3.45	2.35 / 3.37	2.31 / 3.30	2.28 / 3.23	2.23 / 3.13	2.18 / 3.05	2.12 / 2.94	2.08 / 2.86	2.04 / 2.77	1.99 / 2.69	1.96 / 2.63	1.92 / 2.56	1.90 / 2.53	1.87 / 2.47	1.85 / 2.44	1.84 / 2.42
21	4.32 / 8.02	3.47 / 5.78	3.07 / 4.87	2.84 / 4.37	2.68 / 4.04	2.57 / 3.81	2.49 / 3.65	2.42 / 3.51	2.37 / 3.40	2.32 / 3.31	2.28 / 3.24	2.25 / 3.17	2.20 / 3.07	2.15 / 2.99	2.09 / 2.88	2.05 / 2.80	2.00 / 2.72	1.96 / 2.63	1.93 / 2.58	1.89 / 2.51	1.87 / 2.47	1.84 / 2.42	1.82 / 2.38	1.81 / 2.36
22	4.30 / 7.94	3.44 / 5.72	3.05 / 4.82	2.82 / 4.31	2.66 / 3.99	2.55 / 3.76	2.47 / 3.59	2.40 / 3.45	2.35 / 3.35	2.30 / 3.26	2.26 / 3.18	2.23 / 3.12	2.18 / 3.02	2.13 / 2.94	2.07 / 2.83	2.03 / 2.75	1.98 / 2.67	1.93 / 2.58	1.91 / 2.53	1.87 / 2.46	1.84 / 2.42	1.81 / 2.37	1.80 / 2.33	1.78 / 2.31
23	4.28 / 7.88	3.42 / 5.66	3.03 / 4.76	2.80 / 4.26	2.64 / 3.94	2.53 / 3.71	2.45 / 3.54	2.38 / 3.41	2.32 / 3.30	2.28 / 3.21	2.24 / 3.14	2.20 / 3.07	2.14 / 2.97	2.10 / 2.89	2.04 / 2.78	2.00 / 2.70	1.96 / 2.62	1.91 / 2.53	1.88 / 2.48	1.84 / 2.41	1.82 / 2.37	1.79 / 2.32	1.77 / 2.28	1.76 / 2.26
24	4.26 / 7.82	3.40 / 5.61	3.01 / 4.72	2.78 / 4.22	2.62 / 3.90	2.51 / 3.67	2.43 / 3.50	2.36 / 3.36	2.30 / 3.25	2.26 / 3.17	2.22 / 3.09	2.18 / 3.03	2.13 / 2.93	2.09 / 2.85	2.02 / 2.74	1.98 / 2.66	1.94 / 2.58	1.89 / 2.49	1.86 / 2.44	1.82 / 2.36	1.80 / 2.33	1.76 / 2.27	1.74 / 2.23	1.73 / 2.21
25	4.24 / 7.77	3.38 / 5.57	2.99 / 4.68	2.76 / 4.18	2.60 / 3.86	2.49 / 3.63	2.41 / 3.46	2.34 / 3.32	2.28 / 3.21	2.24 / 3.13	2.20 / 3.05	2.16 / 2.99	2.11 / 2.89	2.06 / 2.81	2.00 / 2.70	1.96 / 2.62	1.92 / 2.54	1.87 / 2.45	1.84 / 2.40	1.80 / 2.32	1.77 / 2.29	1.74 / 2.23	1.72 / 2.19	1.71 / 2.17
26	4.22 / 7.72	3.37 / 5.53	2.98 / 4.64	2.74 / 4.14	2.59 / 3.82	2.47 / 3.59	2.39 / 3.42	2.32 / 3.29	2.27 / 3.17	2.22 / 3.09	2.18 / 3.02	2.15 / 2.96	2.10 / 2.86	2.05 / 2.77	1.99 / 2.66	1.95 / 2.58	1.90 / 2.50	1.85 / 2.41	1.82 / 2.36	1.78 / 2.28	1.76 / 2.25	1.72 / 2.19	1.70 / 2.15	1.69 / 2.13

Table F (*Continued*)

n_1 degrees of freedom (for greater mean square)

n_2	1	2	3	4	5	6	7	8	9	10	11	12	14	16	20	24	30	40	50	75	100	200	500	∞
27	4.21 / 7.68	3.35 / 5.49	2.96 / 4.60	2.73 / 4.11	2.57 / 3.79	2.46 / 3.56	2.37 / 3.39	2.30 / 3.26	2.25 / 3.14	2.20 / 3.06	2.16 / 2.98	2.13 / 2.93	2.08 / 2.83	2.03 / 2.74	1.97 / 2.63	1.93 / 2.55	1.88 / 2.47	1.84 / 2.38	1.80 / 2.33	1.76 / 2.25	1.74 / 2.21	1.71 / 2.16	1.68 / 2.12	1.67 / 2.10
28	4.20 / 7.64	3.34 / 5.45	2.95 / 4.57	2.71 / 4.07	2.56 / 3.76	2.44 / 3.53	2.36 / 3.36	2.29 / 3.23	2.24 / 3.11	2.19 / 3.03	2.15 / 2.95	2.12 / 2.90	2.06 / 2.80	2.02 / 2.71	1.96 / 2.60	1.91 / 2.52	1.87 / 2.44	1.81 / 2.35	1.78 / 2.30	1.75 / 2.22	1.72 / 2.18	1.69 / 2.13	1.67 / 2.09	1.65 / 2.06
29	4.18 / 7.60	3.33 / 5.42	2.93 / 4.54	2.70 / 4.04	2.54 / 3.73	2.43 / 3.50	2.35 / 3.33	2.28 / 3.20	2.22 / 3.08	2.18 / 3.00	2.14 / 2.92	2.10 / 2.87	2.05 / 2.77	2.00 / 2.68	1.94 / 2.57	1.90 / 2.49	1.85 / 2.41	1.80 / 2.32	1.77 / 2.27	1.73 / 2.19	1.71 / 2.15	1.68 / 2.10	1.65 / 2.06	1.64 / 2.03
30	4.17 / 7.56	3.32 / 5.39	2.92 / 4.51	2.69 / 4.02	2.53 / 3.70	2.42 / 3.47	2.34 / 3.30	2.27 / 3.17	2.21 / 3.06	2.16 / 2.98	2.12 / 2.90	2.09 / 2.84	2.04 / 2.74	1.99 / 2.66	1.93 / 2.55	1.89 / 2.47	1.84 / 2.38	1.79 / 2.29	1.76 / 2.24	1.72 / 2.16	1.69 / 2.13	1.66 / 2.07	1.64 / 2.03	1.62 / 2.01
32	4.15 / 7.50	3.30 / 5.34	2.90 / 4.46	2.67 / 3.97	2.51 / 3.66	2.40 / 3.42	2.32 / 3.25	2.25 / 3.12	2.19 / 3.01	2.14 / 2.94	2.10 / 2.86	2.07 / 2.80	2.02 / 2.70	1.97 / 2.62	1.91 / 2.51	1.86 / 2.42	1.82 / 2.34	1.76 / 2.25	1.74 / 2.20	1.69 / 2.12	1.67 / 2.08	1.64 / 2.02	1.61 / 1.98	1.59 / 1.96
34	4.13 / 7.44	3.28 / 5.29	2.88 / 4.42	2.65 / 3.93	2.49 / 3.61	2.38 / 3.38	2.30 / 3.21	2.23 / 3.08	2.17 / 2.97	2.12 / 2.89	2.08 / 2.82	2.05 / 2.76	2.00 / 2.66	1.95 / 2.58	1.89 / 2.47	1.84 / 2.38	1.80 / 2.30	1.74 / 2.21	1.71 / 2.15	1.67 / 2.08	1.64 / 2.04	1.61 / 1.98	1.59 / 1.94	1.57 / 1.91
36	4.11 / 7.39	3.26 / 5.25	2.86 / 4.38	2.63 / 3.89	2.48 / 3.58	2.36 / 3.35	2.28 / 3.18	2.21 / 3.04	2.15 / 2.94	2.10 / 2.86	2.06 / 2.78	2.03 / 2.72	1.98 / 2.62	1.93 / 2.54	1.87 / 2.43	1.82 / 2.35	1.78 / 2.26	1.72 / 2.17	1.69 / 2.12	1.65 / 2.04	1.62 / 2.00	1.59 / 1.94	1.56 / 1.90	1.55 / 1.87
38	4.10 / 7.35	3.25 / 5.21	2.85 / 4.34	2.62 / 3.86	2.46 / 3.54	2.35 / 3.32	2.26 / 3.15	2.19 / 3.02	2.14 / 2.91	2.09 / 2.82	2.05 / 2.75	2.02 / 2.69	1.96 / 2.59	1.92 / 2.51	1.85 / 2.40	1.80 / 2.32	1.76 / 2.22	1.71 / 2.14	1.67 / 2.08	1.63 / 2.00	1.60 / 1.97	1.57 / 1.90	1.54 / 1.86	1.53 / 1.84
40	4.08 / 7.31	3.23 / 5.18	2.84 / 4.31	2.61 / 3.83	2.45 / 3.51	2.34 / 3.29	2.25 / 3.12	2.18 / 2.99	2.12 / 2.88	2.07 / 2.80	2.04 / 2.73	2.00 / 2.66	1.95 / 2.56	1.90 / 2.49	1.84 / 2.37	1.79 / 2.29	1.74 / 2.20	1.69 / 2.11	1.66 / 2.05	1.61 / 1.97	1.59 / 1.94	1.55 / 1.88	1.53 / 1.84	1.51 / 1.81
42	4.07 / 7.27	3.22 / 5.15	2.83 / 4.29	2.59 / 3.80	2.44 / 3.49	2.32 / 3.26	2.24 / 3.10	2.17 / 2.96	2.11 / 2.86	2.06 / 2.77	2.02 / 2.70	1.99 / 2.64	1.94 / 2.54	1.89 / 2.46	1.82 / 2.35	1.78 / 2.26	1.73 / 2.17	1.68 / 2.08	1.64 / 2.02	1.60 / 1.94	1.57 / 1.91	1.54 / 1.85	1.51 / 1.80	1.49 / 1.78
44	4.06 / 7.24	3.21 / 5.12	2.82 / 4.26	2.58 / 3.78	2.43 / 3.46	2.31 / 3.24	2.23 / 3.07	2.16 / 2.94	2.10 / 2.84	2.05 / 2.75	2.01 / 2.68	1.98 / 2.62	1.92 / 2.52	1.88 / 2.44	1.81 / 2.32	1.76 / 2.24	1.72 / 2.15	1.66 / 2.06	1.63 / 2.00	1.58 / 1.92	1.56 / 1.88	1.52 / 1.82	1.50 / 1.78	1.48 / 1.75
46	4.05 / 7.21	3.20 / 5.10	2.81 / 4.24	2.57 / 3.76	2.42 / 3.44	2.30 / 3.22	2.22 / 3.05	2.14 / 2.92	2.09 / 2.82	2.04 / 2.73	2.00 / 2.66	1.97 / 2.60	1.91 / 2.50	1.87 / 2.42	1.80 / 2.30	1.75 / 2.22	1.71 / 2.13	1.65 / 2.04	1.62 / 1.98	1.57 / 1.90	1.54 / 1.86	1.51 / 1.80	1.48 / 1.76	1.46 / 1.72
48	4.04 / 7.19	3.19 / 5.08	2.80 / 4.22	2.56 / 3.74	2.41 / 3.42	2.30 / 3.20	2.21 / 3.04	2.14 / 2.90	2.08 / 2.80	2.03 / 2.71	1.99 / 2.64	1.96 / 2.58	1.90 / 2.48	1.86 / 2.40	1.79 / 2.28	1.74 / 2.20	1.70 / 2.11	1.64 / 2.02	1.61 / 1.96	1.56 / 1.88	1.53 / 1.84	1.50 / 1.78	1.47 / 1.73	1.45 / 1.70

Table F (*Continued*)

n_1 degrees of freedom (for greater mean square)

n_2	1	2	3	4	5	6	7	8	9	10	11	12	14	16	20	24	30	40	50	75	100	200	500	∞
50	4.03 7.17	3.18 5.06	2.79 4.20	2.56 3.72	2.40 3.41	2.29 3.18	2.20 3.02	2.13 2.88	2.07 2.78	2.02 2.70	1.98 2.62	1.95 2.56	1.90 2.46	1.85 2.39	1.78 2.26	1.74 2.18	1.69 2.10	1.63 2.00	1.60 1.94	1.55 1.86	1.52 1.82	1.48 1.76	1.46 1.71	1.44 1.68
55	4.02 7.12	3.17 5.01	2.78 4.16	2.54 3.68	2.38 3.37	2.27 3.15	2.18 2.98	2.11 2.85	2.05 2.75	2.00 2.66	1.97 2.59	1.93 2.53	1.88 2.43	1.83 2.35	1.76 2.23	1.72 2.15	1.67 2.06	1.61 1.96	1.58 1.90	1.52 1.82	1.50 1.78	1.46 1.71	1.43 1.66	1.41 1.64
60	4.00 7.08	3.15 4.98	2.76 4.13	2.52 3.65	2.37 3.34	2.25 3.12	2.17 2.95	2.10 2.82	2.04 2.72	1.99 2.63	1.95 2.56	1.92 2.50	1.86 2.40	1.81 2.32	1.75 2.20	1.70 2.12	1.65 2.03	1.59 1.93	1.56 1.87	1.50 1.79	1.48 1.74	1.44 1.68	1.41 1.63	1.39 1.60
65	3.99 7.04	3.14 4.95	2.75 4.10	2.51 3.62	2.36 3.31	2.24 3.09	2.15 2.93	2.08 2.79	2.02 2.70	1.98 2.61	1.94 2.54	1.90 2.47	1.85 2.37	1.80 2.30	1.73 2.18	1.68 2.09	1.63 2.00	1.57 1.90	1.54 1.84	1.49 1.76	1.46 1.71	1.42 1.64	1.39 1.60	1.37 1.56
70	3.98 7.01	3.13 4.92	2.74 4.08	2.50 3.60	2.35 3.29	2.23 3.07	2.14 2.91	2.07 2.77	2.01 2.67	1.97 2.59	1.93 2.51	1.89 2.45	1.84 2.35	1.79 2.28	1.72 2.15	1.67 2.07	1.62 1.98	1.56 1.88	1.53 1.82	1.47 1.74	1.45 1.69	1.40 1.62	1.37 1.56	1.35 1.53
80	3.96 6.96	3.11 4.88	2.72 4.04	2.48 3.56	2.33 3.25	2.21 3.04	2.12 2.87	2.05 2.74	1.99 2.64	1.95 2.55	1.91 2.48	1.88 2.41	1.82 2.32	1.77 2.24	1.70 2.11	1.65 2.03	1.60 1.94	1.54 1.84	1.51 1.78	1.45 1.70	1.42 1.65	1.38 1.57	1.35 1.52	1.32 1.49
100	3.94 6.90	3.09 4.82	2.70 3.98	2.46 3.51	2.30 3.20	2.19 2.99	2.10 2.82	2.03 2.69	1.97 2.59	1.92 2.51	1.88 2.43	1.85 2.36	1.79 2.26	1.75 2.19	1.68 2.06	1.63 1.98	1.57 1.89	1.51 1.79	1.48 1.73	1.42 1.64	1.39 1.59	1.34 1.51	1.30 1.46	1.28 1.43
125	3.92 6.84	3.07 4.78	2.68 3.94	2.44 3.47	2.29 3.17	2.17 2.95	2.08 2.79	2.01 2.65	1.95 2.56	1.90 2.47	1.86 2.40	1.83 2.33	1.77 2.23	1.72 2.15	1.65 2.03	1.60 1.94	1.55 1.85	1.49 1.75	1.45 1.68	1.39 1.59	1.36 1.54	1.31 1.46	1.27 1.40	1.25 1.37
150	3.91 6.81	3.06 4.75	2.67 3.91	2.43 3.44	2.27 3.14	2.16 2.92	2.07 2.76	2.00 2.62	1.94 2.53	1.89 2.44	1.85 2.37	1.82 2.30	1.76 2.20	1.71 2.12	1.64 2.00	1.59 1.91	1.54 1.83	1.47 1.72	1.44 1.66	1.37 1.56	1.34 1.51	1.29 1.43	1.25 1.37	1.22 1.33
200	3.89 6.76	3.04 4.71	2.65 3.88	2.41 3.41	2.26 3.11	2.14 2.90	2.05 2.73	1.98 2.60	1.92 2.50	1.87 2.41	1.83 2.34	1.80 2.28	1.74 2.17	1.69 2.09	1.62 1.97	1.57 1.88	1.52 1.79	1.45 1.69	1.42 1.62	1.35 1.53	1.32 1.48	1.26 1.39	1.22 1.33	1.19 1.28
400	3.86 6.70	3.02 4.66	2.62 3.83	2.39 3.36	2.23 3.06	2.12 2.85	2.03 2.69	1.96 2.55	1.90 2.46	1.85 2.37	1.81 2.29	1.78 2.23	1.72 2.12	1.67 2.04	1.60 1.92	1.54 1.84	1.49 1.74	1.42 1.64	1.38 1.57	1.32 1.47	1.28 1.42	1.22 1.32	1.16 1.24	1.13 1.19
1000	3.85 6.66	3.00 4.62	2.61 3.80	2.38 3.34	2.22 3.04	2.10 2.82	2.02 2.66	1.95 2.53	1.89 2.43	1.84 2.34	1.80 2.26	1.76 2.20	1.70 2.09	1.65 2.01	1.58 1.89	1.53 1.81	1.47 1.71	1.41 1.61	1.36 1.54	1.30 1.44	1.26 1.38	1.19 1.28	1.13 1.19	1.08 1.11
∞	3.84 6.64	2.99 4.60	2.60 3.78	2.37 3.32	2.21 3.02	2.09 2.80	2.01 2.64	1.94 2.51	1.88 2.41	1.83 2.32	1.79 2.24	1.75 2.18	1.69 2.07	1.64 1.99	1.57 1.87	1.52 1.79	1.46 1.69	1.40 1.59	1.35 1.52	1.28 1.41	1.24 1.36	1.17 1.25	1.11 1.15	1.00 1.00

Table G. *Critical values of chi-square**

	Level of significance for one-tailed test					
	.10	.05	.025	.01	.005	.0005
	Level of significance for two-tailed test					
df**	.20	.10	.05	.02	.01	.001
1	1.64	2.71	3.84	5.41	6.63	10.83
2	3.22	4.61	5.99	7.82	9.21	13.82
3	4.64	6.25	7.82	9.84	11.34	16.27
4	5.99	7.78	9.49	11.67	13.28	18.46
5	7.29	9.24	11.07	13.39	15.09	20.52
6	8.56	10.64	12.59	15.03	16.81	22.46
7	9.80	12.02	14.07	16.62	18.48	24.32
8	11.03	13.36	15.51	18.17	20.09	26.12
9	12.24	14.68	16.92	19.68	21.67	27.88
10	13.44	15.99	18.31	21.16	23.21	29.59
11	14.63	17.28	19.68	22.62	24.72	31.26
12	15.81	18.55	21.03	24.05	26.22	32.91
13	16.98	19.81	22.36	25.47	27.69	34.53
14	18.15	21.06	23.68	26.87	29.14	36.12
15	19.31	22.31	25.00	28.26	30.58	37.70
16	20.46	23.54	26.30	29.63	32.00	39.25
17	21.62	24.77	27.59	31.00	33.41	40.79
18	22.76	25.99	28.87	32.35	34.81	42.31
19	23.90	27.20	30.14	33.69	36.19	43.82
20	25.04	28.41	31.41	35.02	37.57	45.32
21	26.17	29.62	32.67	36.34	38.93	46.80
22	27.30	30.81	33.92	37.66	40.29	48.27
23	28.43	32.01	35.17	38.97	41.64	49.73
24	29.55	33.20	36.42	40.27	42.98	51.18
25	30.68	34.38	37.65	41.57	44.31	52.62
26	31.80	35.56	38.89	42.86	45.64	54.05
27	32.91	36.74	40.11	44.14	46.96	55.48
28	34.03	37.92	41.34	45.42	48.28	56.89
29	35.14	39.09	42.56	46.69	49.59	58.30
30	36.25	40.26	43.77	47.96	50.89	59.70

* From R. A. Fisher and F. Yates, "Statistical Tables for Biological, Agricultural and Medical Research," 6th ed. Oliver and Boyd, Edinburgh, 1963. Reproduced by permission of the authors and publishers.

** For *df* greater than 30, the value obtained from the expression $\sqrt{2\chi^2} - \sqrt{2df - 1}$ may be used as a *t* ratio.

Table H. *Power as a function of δ and significance criterion (α)*

δ	One-tailed test (α) .05 / Two-tailed test (α) .10	.025 / .05	.01 / .02	.005 / .01
0.0	.10*	.05*	.02	.01
0.1	.10*	.05*	.02	.01
0.2	.11*	.05	.02	.01
0.3	.12*	.06	.03	.01
0.4	.13*	.07	.03	.01
0.5	.14	.08	.03	.02
0.6	.16	.09	.04	.02
0.7	.18	.11	.05	.03
0.8	.21	.13	.06	.04
0.9	.23	.15	.08	.05
1.0	.26	.17	.09	.06
1.1	.30	.20	.11	.07
1.2	.33	.22	.13	.08
1.3	.37	.26	.15	.10
1.4	.40	.29	.18	.12
1.5	.44	.32	.20	.14
1.6	.48	.36	.23	.16
1.7	.52	.40	.27	.19
1.8	.56	.44	.30	.22
1.9	.60	.48	.33	.25
2.0	.64	.52	.37	.28
2.1	.68	.56	.41	.32
2.2	.71	.59	.45	.35
2.3	.74	.63	.49	.39
2.4	.77	.67	.53	.43
2.5	.80	.71	.57	.47
2.6	.83	.74	.61	.51

δ	One-tailed test (α) .05 / Two-tailed test (α) .10	.025 / .05	.01 / .02	.005 / .01
2.7	.85	.77	.65	.55
2.8	.88	.80	.68	.59
2.9	.90	.83	.72	.63
3.0	.91	.85	.75	.66
3.1	.93	.87	.78	.70
3.2	.94	.89	.81	.73
3.3	.96	.91	.83	.77
3.4	.96	.93	.86	.80
3.5	.97	.94	.88	.82
3.6	.97	.95	.90	.85
3.7	.98	.96	.92	.87
3.8	.98	.97	.93	.89
3.9	.99	.97	.94	.91
4.0	.99	.98	.95	.92
4.1	.99	.98	.96	.94
4.2	.99	.99	.97	.95
4.3	**	.99	.98	.96
4.4		.99	.98	.97
4.5		.99	.99	.97
4.6		**	.99	.98
4.7			.99	.98
4.8			.99	.99
4.9			.99	.99
5.0			**	.99
5.1				.99
5.2				**

* Values inaccurate for *one-tailed* test by more than .01.
** The power at and below this point is greater than .995.

Table I. δ *as a function of significance criterion* (a) *and power*

Power	One-tailed test (a)			
	.05	.025	.01	.005
	Two-tailed test (a)			
	.10	.05	.02	.01
.25	0.97	1.29	1.65	1.90
.50	1.64	1.96	2.33	2.58
.60	1.90	2.21	2.58	2.83
.67	2.08	2.39	2.76	3.01
.70	2.17	2.48	2.85	3.10
.75	2.32	2.63	3.00	3.25
.80	2.49	2.80	3.17	3.42
.85	2.68	3.00	3.36	3.61
.90	2.93	3.24	3.61	3.86
.95	3.29	3.60	3.97	4.22
.99	3.97	4.29	4.65	4.90
.999	4.37	5.05	5.42	5.67

index